Carbon Responsibility and Embodied Emissions

Climate change policy and the reduction of greenhouse gas emissions are currently discussed at all levels, ranging from the Kyoto Protocol to the advertisement of 'carbon neutrality' in consumer products. However, the only policy option usually considered is the reduction of direct emissions. Another potential policy tool, currently neglected, is the reduction of indirect emissions, i.e. the emissions embodied in goods and services, or the payments thereof.

This book addresses the accounting of indirect carbon emissions (as embodied in international trade) within the framework of input–output analysis and derives an indicator of environmental responsibility as the average of consumer and producer responsibility. A global multi-regional input–output model is built, using databases on international trade and greenhouse gas emissions, from which embodied carbon emissions and carbon responsibilities are obtained.

Carbon Responsibility and Embodied Emissions consists of a theoretical part concerning the choice of environmental indicators, and an applied part reporting an environmental multi-regional input–output model. It will be of particular interest to postgraduate students and researchers in Ecological Economics, Environmental Input–Output Analysis and Industrial Ecology.

João Rodrigues is currently an Assistant Researcher at the Center for Innovation, Technology and Policy Research (IN+), Instituto Superior Técnico (IST), Lisbon, Portugal. **Alexandra Marques** is currently an MIT Portugal Program PhD student at IST in the area of Sustainable Energy Systems. **Tiago Domingos** is an Assistant Professor at the Environment and Energy Scientific Area, DEM, IST and a Researcher at IN+.

Routledge Studies in Ecological Economics

Sustainability Networks
Cognitive tools for expert collaboration in social-ecological systems
Janne Hukkinen

Drivers of Environmental Change in Uplands
Aletta Bonn, Tim Allot, Klaus Hubaceck and Jon Stewart

Resilience, Reciprocity and Ecological Economics
Northwest coast sustainability
Ronald Trosper

Environment and Employment
A reconciliation
Philip Lawn

Philosophical Basics of Ecology and Economy
Malte Faber and Reiner Manstetten

Carbon Responsibility and Embodied Emissions
Theory and measurement
João Rodrigues, Alexandra Marques and Tiago Domingos

Carbon Responsibility and Embodied Emissions

Theory and measurement

João Rodrigues,
Alexandra Marques
and Tiago Domingos

LONDON AND NEW YORK

First published 2010 by Routledge
2 Park Square, Milton Park, Abingdon, Oxon, OX14 4RN

Simultaneously published in the USA and Canada
by Routledge
711 Third Avenue, New York, NY10017

*Routledge is an imprint of the Taylor & Francis Group, an informa
business*

First issued in paperback 2011

© 2010 João Rodrigues, Alexandra Marques and
Tiago Domingos

Typeset in Times New Roman by Glyph International Ltd.

British Library Cataloguing in Publication Data
A catalogue record for this book is available from the British
Library

Library of Congress Cataloging in Publication Data
Rodrigues, João F. D.
Carbon responsibility and embodied emissions: theory and
measurement / João F.D. Rodrigues, Alexandra P.S. Marques, and
Tiago M.D. Domingos.
    p. cm.
Includes bibliographical references and index.
1. Carbon dioxide mitigation–Government policy. 2. Atmospheric
carbon dioxide–Measurement. 3. Environmental auditing.
4. Environmental responsibility. I. Marques, Alexandra P. S.
II. Domingos, Tiago M. D. IE. Title. TD885.5.C3.R646 2010
363.738'74-dc22   2009037233

ISBN13: 978-0-415-47020-9 (hbk)
ISBN13: 978-0-203-85574-4 (ebk)
ISBN13: 978-0-415-51864-6 (pbk)

Contents

List of figures	ix
List of tables	xi
Preface	xiii
Acknowledgements	xv

1 Introduction — 1
1.1 *Motivation* — *1*
1.2 *What, why and who* — *3*
1.3 *Overview* — *5*

2 Accounting indirect emissions — 6
2.1 *Input–Output analysis* — *6*
2.2 *Indirect emissions* — *9*
2.3 *Other methods* — *20*

3 Carbon indicators — 22
3.1 *Review* — *22*
3.2 *Comparison* — *29*

4 Carbon responsibility — 33
4.1 *Properties* — *33*
4.2 *Derivation* — *39*

5 Multi-regional IO model — 47
5.1 *Environmental MRIO models* — *47*
5.2 *Data* — *48*
5.3 *Computation* — *52*

6   Carbon responsibility of world regions   56
   6.1   Role of international trade   *56*
   6.2   Responsibilities vs GDP   *60*
   6.3   Responsibilities vs direct emissions   *64*
   6.4   Carbon trade balance   *69*
   6.5   Total carbon responsibility   *74*

7   Discussion   78
   7.1   Summary   *78*
   7.2   Open questions   *79*
   7.3   Embodied emissions   *80*
   7.4   Carbon responsibility   *82*

Appendix A   86
   A.1   List of symbols   *86*
   A.2   GTAP sector and region codes   *87*
   A.3   Carbon responsibilities of GTAP regions   *92*

Bibliography   *100*
Index   *107*

Figures

2.1	3-sector supply chain	13
2.2	Branching chain	14
5.1	GTAP structure	49
6.1	Indirect effects in total carbon intensities	58
6.2	International indirect effects in total carbon intensities	59
6.3	Estimation errors in total carbon intensities	60
6.4	Direct emissions vs GDP	61
6.5	Consumer responsibility vs GDP	62
6.6	Producer responsibility vs GDP	63
6.7	Absolute carbon responsibilities of selected regions	65
6.8	Per capita carbon responsibilities of selected regions	66
6.9	Absolute carbon responsibilities of aggregate regions	68
6.10	Per capita carbon responsibilities of aggregate regions	68
6.11	UCTB vs DCTB for GTAP regions	71
6.12	UCTB vs DCTB for world regions	72
6.13	Total carbon responsibility for GTAP regions	75
6.14	Total carbon responsibility for world regions	76

Tables

6.1	World region elasticities	63
6.2	Absolute responsibilities of aggregated regions	67
6.3	Per capita responsibilities of aggregated regions	67
6.4	Carbon trade of world regions	71
6.5	Intensities of aggregated regions	73
6.6	Total carbon responsibility of world regions	76
A.1	Frequent symbols	86
A.2	GTAP sectors	87
A.3	GTAP regions	89
A.4	GTAP composite regions	90
A.5	World regions	91
A.6	Absolute responsibilities of GTAP regions	92
A.7	Per capita responsibilities of GTAP regions	94
A.8	Carbon trade balance of GTAP regions	97
A.9	Total carbon responsibility of GTAP regions	98

Preface

Climate change policy and the reduction of greenhouse gas emissions are currently discussed at all scales, ranging from the Kyoto Protocol to the increasingly frequent advertisement of 'carbon neutrality' in consumer products. However, the only policy option usually considered is the reduction in direct emissions in several forms. Another potential policy tool, currently neglected, is the reduction of indirect emissions, i.e. the emissions embodied in goods and services, or the payments thereof.

In this book we report the mathematical technique for computing the total carbon emissions embodied in goods and services and the payments thereof, and derive an indicator of carbon responsibility which is applied to countries (or other economic agents).

We emphasize the possibility of abating both upstream and downstream embodied emissions as a means to achieve reductions in direct carbon emissions.

We use the Global Trade Analysis Project (GTAP) 6 database (87 world regions, 57 sectors per region, base year 2001) to build a global multi-regional input–output (MRIO) model and compute the embodied emissions of economic sectors and the carbon responsibility of regions.

Among other results, we observe that with regard to economic and environmental performance, countries can be grouped in world regions. For Developed economies, Africa and Latin America, carbon responsibility is higher than direct emissions, whereas for Fossil Fuel Exporters, Asia and Eastern Europe, the opposite is true. When considering individual countries, greater variability is found, and especially for small open economies carbon responsibility can deviate substantially from direct emissions.

Acknowledgements

The work reported here benefited from discussions with many people, including Tânia Sousa, Ana Simões, Rui Mota, Stefan Giljum, François Schneider, Manfred Lenzen, Thomas Wiedmann, Robbie Andrew and Glen Peters.

We particularly acknowledge the help of Glen Peters in the processing of the GTAP data, of João Magalhães and Rosa Trancoso in programming, and of Natalia Petrenko and Kamila Krakowska in correcting the text.

We would like to acknowledge the financial support of FCT through scholarship SFRH/BPD/36136/2007 (to JR), scholarship SFRH/BD/42491/2007 (to AM) and grant PTDC/AMB/64762/2006.

Por fim, agradeçemos às nossas famílias e amigos. Pelo apoio constante, um agradecimento profundo.

1   Introduction

1.1 Motivation

1.1.1 Climate change policy

According to the International Panel on Climate Change (IPCC),

> [t]he global atmospheric concentration of carbon dioxide has increased
> from a pre-industrial value of about 280 ppm to 379 ppm3 in 2005,
> while the global surface temperature rose $0.74 \pm 0.18°$ C and [t]he linear
> warming trend over the last 50 years ($0.13 \pm 0.03°$ C per decade) is
> nearly twice that for the last 100 years.
>
> (IPCC 2007)

In the first decade of the twenty-first century, in spite of some dissenting
voices (Schulte 2008), a consensus emerged among scientists, policy-
makers and the general public that climate change is taking place and that
the main drivers of global warming are anthropogenic, the most important
of them being the emission of greenhouse gases (GHG) from the burning of
fossil fuels (Canadell *et al.* 2007, Hansen and Sato 2004, IPCC 2007).

The evidence of the anthropogenic role in global warming was mounting in
the last decades of the twentieth century and, under the auspices of the United
Nations, the United Nations Framework Convention on Climate Change
(UNFCCC) was formed to coordinate efforts to mitigate global warming on
a global scale. Under the UNFCCC, the Kyoto Protocol (UNFCCC 1998)
was adopted in 1997, entered into force in 2005 and is due to expire in 2012.

> Recognizing that developed countries are principally responsible for
> the current high levels of GHG emissions in the atmosphere as a result
> of more than 150 years of industrial activity, the Protocol places a

heavier burden on developed nations under the principle of 'common but differentiated responsibilities'.

(http://unfccc.int/kyoto_protocol/items/2830.php)

Thus, the Protocol commits developed countries to stabilize GHG emissions, while developing countries are only encouraged to do so.

1.1.2 Policy scales

Currently, initiatives to curb global warming are being implemented at all levels. On a national and global scale, there is the Kyoto Protocol, with its mechanisms of emissions trading, clean development mechanism (CDM) and joint implementation (JI). In order to cope with the Kyoto Protocol, the European Union established a cap-and-trade system the European Trading Scheme (ETS) for large emitters in the energy and industrial sectors (Grubb and Neuhoff 2006). Through the ETS, firms which don't use all their permits can sell them to other firms. Several whole countries (Iceland, Norway, Costa Rica and New Zealand) are even aiming to become carbon neutral in the coming decades (Olafsson 2008).

On the smaller scale of specific sectors, the official policies being undertaken are too many to be enumerated. However, it is not just official institutions that are engaged in climate change policy. Some firms voluntarily report GHG inventories (WBCSD 2004) and are actively engaged in the reduction of their GHG emissions (Heal 2005), sometimes using ecological labels (Economist 2008, Muller 2007). There is currently a strong interest in the carbon footprint (EPLCA 2008) and the British patent office is advancing towards the publication of a standard for the quantification of GHG emissions embodied in products (BSI 2008).

Besides official institutions and firms, some consumers participate in climate change policy by neutralizing their carbon emissions (for example of airflights) via carbon offsetting schemes (Murray and Dey 2007).

1.1.3 Indirect emissions

The brief review in Section 1.1.2 shows that there is a genuine interest among policy levels, from national governments to firms and consumers, in reducing GHG emissions. These abatement schemes involve a mix of reductions in direct and indirect emissions.

At one extreme, the inventories of GHG emissions, used to assess the compliance of Annex I countries, are compiled on a territorial basis (IPCC 2007). At an intermediate level, the standard for corporate GHG accounting (WBCSD 2004) considers both direct emissions and indirect emissions associated with energy consumption. At the other extreme, the carbon footprint of products (ideally) accounts for all indirect emissions (BSI 2008, EPLCA 2008).

The accounting of indirect emissions poses both theoretical problems (because several indicators exist) and methodological problems (because they involve uncertainty and require more data and computation effort than direct emissions). However, the accounting of indirect emissions is necessary to the making of environmentally efficient policy decisions.

A good example of the importance of indirect emissions is the recent debate surrounding biofuels. Biofuels were proposed as a renewable energy source, because carbon burned from a season's harvest is again fixated, through photosynthesis, in the following season. However, the indirect GHG emissions associated with cultivation, harvesting, processing and land use change can, in some cases, lead to positive net emissions (Worldwatch Institute 2007).

1.2 What, why and who

1.2.1 What this book is about

This book is about the accounting of indirect carbon emissions. We are primarily interested in embodied carbon emissions, which are defined for economic flows (goods and services), and in carbon responsibility, which is defined for economic agents (such as countries or firms).

Embodied carbon emissions are a measure of the total GHG emissions occurring throughout the life cycle of products (EPLCA 2008). We shall distinguish upstream embodied emissions – those required to generate the product itself or occurring upstream in the product's life cycle – and downstream embodied emissions – those required to generate the payment of the product or occurring downstream in the product's life cycle.

Following a literature trend that began a decade ago (Kondo *et al.* 1998), we shall use the term 'carbon responsibility' to denote an indicator that assigns indirect emissions to an economic agent. Carbon responsibility is constructed using embodied emissions (Rodrigues and Domingos 2008), and is a measure of an agent's share in total GHG emissions.

In the first part we shall address the theoretical problem of defining embodied emissions and carbon responsibility. In the second part we shall address the methodological problem of measuring them.

There are two complementary approaches for the accounting of embodied emissions: life-cycle analysis (LCA) (ISO 2006) and input–output (IO) analysis (UN 1999). Our focus will be on the latter technique, although our results also apply to the former.

We will attempt primarily to present rigorous results (within the limits of our abilities and data availability), but at some points we will indulge in some informal considerations (for example regarding policy implications and accounting uncertainties).

We will also present mainly the results of our own work. We review the work of others in the field but do not pretend to present a systematic review.

1.2.2 Why does it matter?

The accounting of embodied emissions and carbon responsibility is important, because these are potential tools for climate change policy, although they currently do not receive much attention. Carbon responsibility could be a useful instrument in environmental negotiations, fostering more involvement than the conventional use of direct emissions (Peters and Hertwich 2008b). Embodied emissions, with appropriate carbon labelling, offer an avenue for indirect abatement through the choice of inputs and outputs (Rodrigues and Domingos 2008).

However, the use of these indicators requires conceptual and methodological clarity as to what they mean and how they are measured. That is, measures of indirect emissions can only gain the trust of economic agents (consumers, firms, governments) if they are clear and accurate.

Currently, there is a lack of clarity in this field, due to a large disparity in terminology and diversity of indicators (e.g. carbon footprint, embedded emissions, embodied emissions).

There are also many uncertainties affecting the results. On the one hand there is a problem of data, as both IO and LCA methods require data that is often fragmentary and outdated. On the other hand it is difficult to estimate the uncertainty of these indicators, as the source data does not often report errors.

Thus with this work we also wish to bring some clarity to the field and raise awareness of the work that remains to be done.

1.2.3 Who should read this book

This book is intended primarily for those working in the fields of environmental IO analysis, LCA, industrial ecology and environmental/ecological economics, or anyone else who is interested in the accounting of embodied emissions and carbon responsibility.

Some knowledge of linear algebra and differential calculus is helpful but not essential for reading the book, as the more mathematical parts are self-contained, and the main results can be followed skipping the technicalities.

1.3 Overview

The main theoretical problem addressed here is which indicator ought to be chosen to define the carbon responsibility of an economic agent. Several such indicators have been proposed in the literature, and our approach to this problem is to define the properties that the indicator should possess and to construct an indicator which satisfies these properties. In this respect, our results are novel, because they draw attention to downstream embodied emissions, which are seldom addressed in the literature.

These theoretical problems are addressed in the first chapters of the book. In Chapter 2 we review the essentials of environmental IO analysis and derive the formulas that compute embodied emissions. In Chapter 3 we review existing carbon indicators, clarifying their practical and ethical implications. In Chapter 4 we discuss the requirements and implications of the properties of environmental indicators, presenting a set of formal properties that define carbon responsibility, and then derive the mathematical expression that defines it.

On the empirical side, we use the GTAP 6 database to present embodied emissions and carbon responsibilities for a world model of 87 regions and 57 sectors for the year 2001. Among other results, we observed that countries can be grouped in world regions, with regard to economic and environmental performance. For developed economies, Africa and Latin America carbon responsibility is higher than direct emissions, whereas for Fossil Fuel Exporters, Asia and Eastern Europe, the opposite is true. When considering individual countries greater variability is found; and for small open economies especially, carbon responsibility can deviate substantially from direct emissions.

These empirical issues are addressed in the last chapters of the book.

In Chapter 5 we move on to the quantification of embodied emissions. We report a multi-regional IO model built from the GTAP database, and the processing and computations involved. In Chapter 6 we report the results of our analysis, distinguishing international estimation errors and international indirect effects.

In Chapter 7 we summarize the main results, discuss policy implications and suggest directions for future research.

In the Appendix we summarize the notation, the GTAP nomenclature and report sectoral carbon intensities.

2 Accounting indirect emissions

2.1 Input–Output Analysis

2.1.1 Economic IO Analysis

Input–output analysis was born from the work of the Economics Nobel prizewinner Wassily Leontief (1905–99). In his early career Leontief wanted to build an empirical description of the economy, without 'theoretical assumptions and unobserved facts'. So, in contrast to the theoretical work that was highly praised at the time, Leontief 'got his hands dirty' with empirical facts, and published in 1941 the *Structure of the American Economy, 1919–29*, which reported the first input–output (IO) table.

An IO table is a matrix whose entries represent the transactions occurring during one year between sectors (the output of a sector is the input of another). All sectors (e.g., agriculture, manufacturing and services) in the economy (e.g., a country, a region, or a city) should be represented. Besides inter-industry transactions, the table also reports transactions involving sectors and final demand, exports, factor payments and imports.

An IO table is a snapshot of the economy, displaying the circular flow of value from households to firms (in the form of labour and capital) and then from firms to households (in the form of goods and services), as well as the circular flow of money from households to firms (goods and service payments) and then from firms to households (in the form of wages and rents). IO tables are regularly collected by national statistical offices (UN 1999).

The main purpose of IO tables, and the one in which we are interested, is to account for indirect effects, e.g., how the increase in the demand of a certain sector affects the supply of another. This can be done using simple linear algebra and the empirical IO table by means of the theoretical *economic* Leontief (or IO) model (Miller and Blair 1985, Raa 2006).

The Leontief model assumes that firms have a production function of constant returns to scale which admits only perfect complements (with zero elasticity of substitution) whose coefficients can be obtained directly from the IO table. For example if the inputs of the car sector are steel and electricity, the production function would be $y = c \min \{ax, bz\}$, where x and z are the inputs, y is the output and a, b and c are coefficients. This is a realistic assumption only in the short run and only to study marginal variations.

In Section 2.2 we shall use the mathematical techniques of IO analysis to quantify indirect carbon emissions. However, even though the calculations in both economic and environmental IO analysis are identical, the interpretation of those calculations is quite different.

2.1.2 Economic vs environmental IO models

Economic and environmental IO models differ in the assumptions involved and in the meaning of the final quantities computed.

An economic IO model (Raa 2006) is a statement about the use of production factors by firms in order to generate output. Therefore the matrix of technical coefficients, which is the set of production recipes used by firms, derived from the IO table, is the mechanistic basis of the model.

From the matrix of technical coefficients, economic IO analysis computes 'multipliers', which are dimensionless quantities that state by how much a certain economic quantity should be multiplied in case something else happens (for example, how much the output of a certain sector increases if the demand for another sector rises by a certain amount).

Since the economic IO model has a mechanistic basis, there are substantial differences between 'demand driven' vs 'supply pushed' models and 'quantity' vs 'price' models (Davar 1989, Dietzenbacher 1997, Kornai 1979, Oosterhaven 1996). (For example in a demand driven model, the production structure is fixed and final demand is exogenously determined.) The practical implication of these differences is that the Ghosh and Leontief inverses (which will be further discussed in Section 2.2) should not be applied to the same data set.

Fortunately for us, environmental IO analysis (Rodrigues *et al.* 2006) is simpler in the assumptions involved and hence less restrictive in the range of application. Environmental IO analysis does not try to explain how the economy works. Instead, an *environmental IO model* is just the partitioning of the economy into sectors and the quantification of economic flows and (for example) direct carbon emissions.

Another important distinction is that instead of economic indirect effects we are concerned with the quantification of environmental (e.g., carbon) indirect effects and therefore the primary quantities that we wish to compute are not multipliers, which are dimensionless, but *intensities*, which have dimensions of carbon emissions over economic value (e.g., $kgCO_2/USD$). A carbon intensity expresses the amount of direct carbon emissions required to generate a specified economic flow.

Given the empirical nature of the environmental model, there is no problem in applying the Ghosh and Leontief inverses to the same data set. As we shall see in Section 2.2, in an environmental context the Ghosh inverse is used to compute upstream indirect effects, i.e. the indirect emissions required to generate a given unit of economic output; and the Leontief inverse is used to compute downstream indirect effects, i.e. the indirect emissions required to generate the payment for a given unit of economic input.

Keeping these conceptual distinctions in mind, we shall apply the mathematical techniques of economic to environmental IO analysis.

2.1.3 Data sources and types of models

The standard procedure for the compilation of IO tables is described in the *System of National Accounts (SNA) 1993* and supplementary material published by the United Nations (UN 1994). We shall use nomenclature and notation close to the SNA 1993 standard as far as possible.

National IO tables are periodically published by national statistical offices, and the current standard is to publish *make* and *use* tables. Such tables do not report transactions between industries but between industries and product types (the make table) and between product types and industries (the use table). Rectangular make and use tables need to be converted to square industry tables in order to compute intensities.

Social accounting matrices (SAMs) are extensions of the IO table that further specify transactions that are not reported in an IO table, such as the taxes paid by the different classes of final demand.

National tables (whether in industry-by-industry or in make use format) are the best documented in the empirical IO literature. Sub-national tables and tables of international transactions (Ackerman *et al.* 2007, Oosterhaven *et al.* 2007) have also been reported, although not in a systematic manner.

Our interest in IO analysis is the quantification of indirect effects; using only a domestic IO table, it is not possible to account for international indirect effects. A multi-regional IO model, in which several regions are represented, ideally covering the whole world, is better at quantifying indirect effects but poses important empirical problems.

National tables differ substantially from one another. Some are large (the US or Chinese tables have several hundred sectors), while others are small (harmonized national EU tables, published by EUROSTAT, have less than 100 sectors). Two common problems affecting the compilation of multi-regional models are that sector classification is not consistent across countries and that national tables are reported in national currencies and thus need to be converted into a common currency.

Another problem concerning multi-regional models is that detailed international inter-industry transactions are, in general, not available and must be estimated from aggregate trade data. However, there exist some harmonized databases that report both national tables and international inter-industry transactions, such as the EUROSTAT (2009), the OECD (Yamano and Ahmad 2006) and GTAP (Dimaranan 2006) databases.

A final problem affecting all types of IO model is that they are static and compiled for a specific reference year. Analysis outside that reference year requires a further update of the table (Wiedmann *et al.* 2008a).

The analysis we develop next applies to an ideal industry-by-industry IO model covering the whole world. With some care in mind, the work can be applied to other situations, such as make use tables, SAMs and national tables.

2.2 Indirect emissions

2.2.1 Environmental IO model

The IO model is as follows. The world is partitioned into a set of N sectors or internal economic agents (we do not distinguish countries, for now). All direct carbon emissions are assigned to a sector and all economic transactions are assigned to a pair of sectors.

We use e_i^L to denote the *direct* (or local) *carbon emissions* of sector i and t_{ij} to denote the economic flow (or transaction) from i to j.

In this book we are interested in computing indirect carbon emissions, where 'carbon,' actually means any greenhouse gas (such as methane or chlorofuorocarbons) converted in units of carbon dioxide equivalent ($kgCO_2$) through the use of greenhouse warming potential (GWP) factors. Emissions are measured in units of mass over time, but as a rule we shall use only mass, implying that those emissions take place during the period of one year.

The analysis developed here can be applied to other environmental indicators (such as emissions of acidifying gases, ecological footprint, water, or energy consumption), but is particularly well suited to pressure indicators (or flow pollutants), rather than state indicators (or stock pollutants) (Eder and Nadoroslawsky 1999), since the latter pose substantial temporal allocation problems; it is also best suited to global, rather than local, environmental problems (Rodrigues *et al.* 2006). So, we shall assign indirect carbon emissions, but not carbon stocks present in the atmosphere, to economic agents.

We consider that there is an external sector, denoted 0, with whom the internal agents (or sectors) have transactions. The total output of a sector equals its total input, as any net difference in transactions with other internal sectors appears as a flow to or from the external sector. We shall use t_i to denote the total input/output of sector i.

Since a transaction is both the output of a sector and the input of another, transactions are linked by the following constraints:

$$t_i = \sum_{j=1}^{N} t_{ij} + t_{i0}, \quad \text{for } i = 1, \dots, N \tag{2.1}$$

$$t_i = \sum_{j=1}^{N} t_{ji} + t_{0i}, \quad \text{for } i = 1, \dots, N \tag{2.2}$$

These quantities (t_{ij}'s, e_i^L's) define the empirical environmental IO model (Section 2.1.2). However, for clarity and convenience, we disaggregate these economic agents and transactions into more easily recognizable quantities.

We now review the nomenclature to be used for the remainder of this book, in accordance with the System of National Accounts.

What we call an *agent* is what SNA 1993 (IV.A.4.2) refers to as an institutional unit: 'an economic entity that is capable, in its own right, of owning assets, incurring liabilities and engaging in economic activities and in transactions with other entities'. Institutional units can be grouped in institutional sectors (SNA 1993, IV.A.4.6) and for the purposes of this book we consider only three institutional sectors: firms, government and households.

We will refer to internal agents indistinctly as firms, industries, or (internal) *sectors*. According to SNA 1993 the firms' sector can be disaggregated into different levels (local units, establishments and industries). In the present work we consider an industry to 'consist of a group of establishments engaged in the same, or similar, kinds of production activity' (V.B.5.5). We also consider that each industry produces a homogeneous *product*, where '[g]oods and services, also called products, are the result of production' (II.B.2.49).

We consider that the world is partitioned into a set of mutually exclusive *regions* (e.g., countries or composite regions such as 'rest of the world'), and that each institutional unit is resident in some region (SNA 1993, IV.A.4.15). The institutional units considered in the first part of this book are N *sectors* and one *external sector*, 0, which encompasses government and households. In the second part of this book the external sector is further disaggregated.

Transactions are economic flows, which 'reflect the creation, transformation, exchange, transfer or extinction of economic value; they involve changes in the volume, composition, or value of an institutional unit's assets and liabilities' (SNA 1993, III.C.3.9). Most (but not all) economic flows are economic transactions, where a 'transaction is an economic flow that is an interaction between institutional units by mutual agreement' (III.C.3.12) and a 'monetary transaction is one in which one institutional unit makes a payment (receives a payment) or incurs a liability (receives an asset) stated in units of currency' (III.C.3.16). We consider all economic flows to be monetary transactions.

According to SNA 1993 (VI.B.6.15) '*production* may be defined as an activity carried out under the control and responsibility of an institutional unit that uses inputs of labour, capital and goods and services to produce outputs of goods or services'. Consumption is an activity in which institutional units use up goods or services and which can be either intermediate or final. *Intermediate consumption* consists of inputs into processes of production.

> Final consumption consists of goods and services used by individual households or the community to satisfy their individual or collective needs or wants. The activity of gross fixed capital formation, on the other hand, is restricted to institutional units in their capacity as producers, being defined as the value of their acquisitions less disposals of fixed assets.
>
> (SNA 1993, I.H.1.49)

Final expenditure consists of final consumption expenditure (by households and government) and gross fixed capital formation. (Thus, acquisition of private households is not gross fixed capital formation, and gross fixed capital formation can be negative.)

In the production account of an industry (SNA 1993, I.B.1.6) gross value added 'is defined as the value of output less the value of intermediate consumption'.

> Primary incomes are incomes that accrue to institutional units as a consequence of their involvement in processes of production or ownership of assets that may be needed for purposes of production. They are payable out of the value added created by production. [...] Receipts from taxes on production and imports are treated as primary incomes.
>
> (SNA 1993, VII.A.7.2)

According to SNA 1993 (VII.A.7.13) primary incomes can be broken down into:

(a) Compensation of employees receivable by households; (b) Taxes (less subsidies) on production or imports receivable (or payable) by government units; (c) Operating surplus, or mixed income, of enterprises [...]; (d) Interest, dividends and similar incomes receivable by the owners of financial assets; (e) Rents receivable by owners of land or sub-soil assets leased to other units.

We shall refer to *final demand* as the sum of all goods and services provided by internal sectors to the external sector, and to *added value* as the sum of all goods and services provided by the external sector to the internal sectors.

That is, we consider the internal sectors to be firms and the external sector to account for households, government and firms the latter in the double role of consumers of fixed assets and receivers of surplus, rents, interest, etc.

This model therefore describes the circular motion in the economy. Value flows from the external sector to firms in the form of added value and from firms to the external sector again in the form of goods and services. Money flows in the opposite direction, from the external sector to firms via final demand, and from firms to the external sector in the form of factor payments.

In the following subsection we describe the linear motion in the economy.

2.2.2 Embodied emissions and carbon intensity

The production and consumption of goods and services requires not only economic transactions but also the absorption and emission of physical flows, such as carbon dioxide and other greenhouse gases.

Thus, a natural and important question is how much do carbon emissions from any particular product 'cost'? Formally, we ask what are the total carbon emissions occurring along the life cycle of a product or, informally, what is its 'carbon footprint'.

This question is important for environmental policy, because it allows the comparison of the environmental performance of products or the quantification of the environmental performance in the successive stages of the life cycle of a particular product.

Total upstream embodied carbon emissions of a flow (TUECEF), e_{ij}^U, is a quantity, with dimensions of direct carbon emissions (e.g., $kgCO_2$), that accounts for all direct and indirect emissions required to generate economic flow (ij).

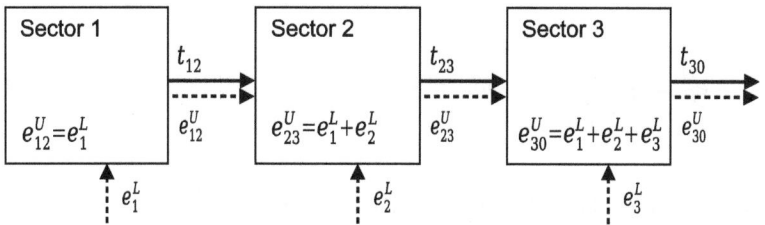

Figure 2.1 Total upstream embodied emissions in a 3-sector supply chain. Full arrows are economic transactions; dashed lines are carbon emissions (direct or embodied).

Let us consider the example of a 3-sector supply chain. Consider a world which has only sectors 1, 2 and 3, with transactions only from 1 to 2, from 2 to 3 and from 3 to 0 (0 is the external sector), with transaction values of $t_{12} = t_{23} = t_{30} = 10$ USD. Furthermore, consider that the direct emissions of the several sectors are $e_1^L = 1$ kgCO$_2$, $e_2^L = 5$ kgCO$_2$ and $e_3^L = 10$ kgCO$_2$. The total upstream embodied emissions of flow 30 is $e_{30}^U = 1 + 5 + 10 = 16$ kgCO$_2$. That is, the delivery of flow 30 (which should be read as 'from 3 to 0') requires the emission of 16 kgCO$_2$. Figure 2.1 shows the supply chain.

If we divide the total upstream embodied carbon emissions of a flow by its monetary value, we obtain another quantity, *total upstream carbon intensity of a flow* TUCIF, which we denote m_{ij}^U, where

$$m_{ij}^U = e_{ij}^U / t_{ij} \tag{2.3}$$

In the case of this supply chain, $m_{30}^U = 16/10 = 1.6$ kgCO$_2$/USD. Carbon intensity has dimensions of emissions per monetary value.

Carbon intensity is an additive quantity, which can be decomposed into direct effects and indirect effects of different orders, where 'order' refers to the distance, in terms of number of sectors, between the flow being considered and the sector where direct emissions are occurring. For example, in the supply chain case considered above, the direct and the 1st and 2nd order indirect effects of flow 30 are, respectively, 1.0, 0.5 and 0.1 kgCO$_2$/USD.

Let us define the auxiliary variable *total upstream embodied carbon emissions of a sector* (TUECES), e_i^U, as $e_i^U = \sum_{j=0}^{N} e_{ij}^U$. TUECES can be computed as the sum of the direct emissions of that sector plus the total

TUECEF of the inputs of that sector:

$$e_i^U = e_i^L + \sum_{j=1}^{N} e_{ji}^U, \quad \text{for } i = 1, \ldots, N \tag{2.4}$$

This formula allows for the recursive computation of e_{ji}^U in the case of a linear supply chain or fusing chains. However, it is ambiguous in the case of a branching chain. For the sake of clarity let us consider a single sector 1, which has $e_1^L = 10 \text{ kgCO}_2$ and delivers flows to sectors 2 and 3, of magnitude $t_{12} = 3$ and $t_{13} = 1$ USD (see Figure 2.2).

The outgoing TUECEFs, e_{12}^U and e_{13}^U must sum up to the sector TUECES, $e_1^U = e_1^L = 10 \text{ kgCO}_2$, but how much of it should be assigned to each? We will discuss other allocation rules, but now we shall follow *economic causality*, which assumes that total upstream embodied emissions are shared in the same proportion as economic output. Flows t_{12} and t_{13} are, respectively, 3/4 and 1/4 of total output, t_1. Therefore, we assign $e_{12}^U = 10\frac{3}{4} = 7.5 \text{ kgC}$ and $e_{13}^U = 10\frac{1}{4} = 2.5 \text{ kgCO}_2$.

We can now compute intensities of all flows (dividing embodied emissions by monetary value) and observe that they match: $m_{12}^U = m_{13}^U = 2.5$ kgCO$_2$/USD. That is, all outputs of a sector have the same TUCIF (Eq. 2.3). Let us define *total upstream carbon intensity of a sector* (TUCIS), m_i^U, as

$$m_i^U = \frac{e_i^U}{t_i} \tag{2.5}$$

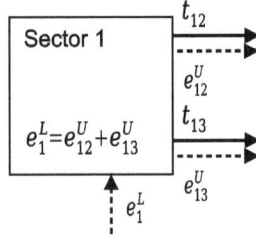

Figure 2.2 Total upstream embodied emissions in a branching chain. Full arrows are economic transactions; dashed lines are carbon emissions (direct or embodied).

Note that this quantity, too, matches the output intensities, $m_1^U = 2.5$ kgCO$_2$/USD.

That is, economic causality solves the ambiguity in the assignment of TUECEFs that arises in branching chains. A mathematical formulation of economic causality is:

$$m_i^U = m_{ij}^U, \quad \text{for } i,j = 1,\ldots,N. \tag{2.6}$$

Equations 2.3, 2.4 and 2.6 jointly define all e_{ij}^U's. It is interesting to observe the emissions embodied in the set of outputs leaving the economy (i.e. final demand), the e_{i0}^U's. Equation 2.4 is conservative, meaning that along a chain all direct emissions are accumulated, and therefore the total upstream embodied emissions in final demand match total direct emissions.

$$\sum_{i=1}^{N} e_{i0}^U = \sum_{i=1}^{N} e_i^L$$

This is a mathematical representation of the linear motion of the economic process, since as one moves along the supply chain, embodied emissions can only accumulate. (Of course, some economic activities lead to carbon abatement or negative emissions, and in the same way as we compute embodied emissions, it is possible to compute embodied abated emissions.)

So far we have addressed only *upstream* embodied emissions, which follow the accumulation of value along a supply chain. However, in every transaction, while the buyer receives the value of a product – and implicitly its upstream embodied emissions – the seller receives the payment for that product – and implicitly its *downstream* emissions. Downstream emissions are not usually studied but, from a mathematical point of view, they are symmetrical to upstream emissions and can be accounted for just as easily. They are computed as:

$$e_i^D = e_i^L + \sum_{j=1}^{N} e_{ij}^D \quad \text{for } i = 1,\ldots,N \tag{2.7}$$

$$m_{ij}^D = e_{ij}^D / t_{ij} \tag{2.8}$$

$$m_i^D = \frac{e_i^D}{t_i} \tag{2.9}$$

$$m_i^D = m_{ji}^D, \quad \text{for } i,j = 1,\ldots,N \tag{2.10}$$

and the same terminology and notation applies as to upstream emissions, except that superscript D is used instead of U. Downstream emissions

(TDECEF and TDECES, TDCIF and TDCIS) accumulate indirect emissions along a demand chain, rather than a supply chain, and total downstream emissions of primary inputs must match total direct emissions:

$$\sum_{i=1}^{N} e_{0i}^{D} = \sum_{i=1}^{S} e_{i}^{L}$$

For any transaction in the economic process, (ij), its TUECEF or e_{ij}^{U} represents the amount of carbon emissions required to physically generate the product, or the carbon footprint of its past life, and its TDECEF e_{ij}^{D} represents the carbon emissions required to generate the payment for that product. Thus, the sum of those two quantities represents the total carbon footprint throughout the life cycle of a product.

In the following subsection we address the methodological problem of computing these quantities in an efficient manner, and in the last Section of this Chapter we mention other definitions of embodied emissions.

2.2.3 Computation

The recursive formulas 2.4 and 2.7 allow the computation of all desired quantities, but such an approach is not feasible for a large system. Let us introduce matrix notation to describe the IO system, and use linear algebra to compute these quantities in a more efficient way (Miller and Blair 1985, Raa 2006, Rodrigues *et al.* 2006, UN 1994 1999). **Bold** denotes matrix or vector (UPPERCASE means matrix, lowercase means vector). $'$ denotes transpose (i.e. $\mathbf{A}'$ is $\mathbf{A}$ with columns and rows interchanged). $\cdot$ denotes the Hadamard (or entrywise) product. $\mathbf{I}$ denotes the identity matrix and $\mathbf{1}$ denotes a vector of ones. Unless stated otherwise, vectors are column vectors.

Let $\mathbf{T}$ be the matrix of *inter-sectoral transactions*, where the entry in the i-th row and j-th column, $T(i,j)$ takes value t_{ij}. Let $\mathbf{y}$ be the vector of *final demand*, where the i-th entry takes value t_{i0}. Let $\mathbf{v}$ be the vector of *added value*, where the i-th entry takes value t_{0i}. Let $\mathbf{x}$ be the vector of *total input/output*, where the i-th entry takes value t_i.

Equations 2.1 and 2.2 can be rewritten as:

$$\mathbf{T1} + \mathbf{y} = \mathbf{x}$$

$$\mathbf{T'1} + \mathbf{v} = \mathbf{x}$$

Let us now introduce $\mathbf{e}$'s and $\mathbf{m}$'s as the vectors of emissions and intensities (direct, L, upstream, U and downstream, D). Direct carbon intensity of sector i, m_i^{L}, is an auxiliary quantity defined as the direct emissions over the total input/output of that sector:

$$m_i^L = \frac{e_i^L}{t_i} \tag{2.11}$$

The upstream and downstream carbon intensities are defined respectively in Eqs 2.3 and 2.6, and 2.8 and 2.10. The components of the vector direct emissions, $\mathbf{e}^L$, are described in the beginning of Section 2.2.1. The components of the vectors of total upstream and downstream embodied emissions, $\mathbf{e}^U$ and $\mathbf{e}^D$, are the TUCESs and TDCESs (total embodied emissions of sectors) described in Section 2.2.2. In the following set of equations we shall also use the matrices of total upstream and downstream embodied emissions, $\mathbf{E}^U$ and $\mathbf{E}^D$, whose elements are the TUCEF and TDCEFs (total embodied emissions of flows), also described in Section 2.2.2.

Recalling the definitions of Subsection 2.2.2, Equations 2.4 and 2.7 can be written in matrix formulation as:

$$\mathbf{e}^U = \mathbf{e}^L + (\mathbf{E}^U)'\mathbf{1}$$

$$\mathbf{e}^D = \mathbf{e}^L + \mathbf{E}^D\mathbf{1}$$

Using expressions 2.5, 2.9 and 2.11, the previous equations can be written as:

$$\mathbf{m}^U \cdot \mathbf{x} = \mathbf{m}^L \cdot \mathbf{x} + \mathbf{T}'\mathbf{m}^U,$$

$$\mathbf{m}^D \cdot \mathbf{x} = \mathbf{m}^L \cdot \mathbf{x} + \mathbf{T}\mathbf{m}^D.$$

Let us define the matrix of 'technical coefficients' as $\mathbf{A}$, such that $A(i,j) = t_{ij}/x_j$. The previous equations can be rewritten as:

$$\mathbf{m}^U \cdot \mathbf{x} = \mathbf{m}^L \cdot \mathbf{x} + \mathbf{A}'(\mathbf{m}^U \cdot \mathbf{x}),$$

$$\mathbf{m}^D \cdot \mathbf{x} = \mathbf{m}^L \cdot \mathbf{x} + \mathbf{A}(\mathbf{m}^D \cdot \mathbf{x}).$$

Eliminating $\mathbf{x}$ from both sides of both equations we obtain:

$$\mathbf{m}^U = \mathbf{m}^L + \mathbf{A}'\mathbf{m}^U,$$

$$\mathbf{m}^D = \mathbf{m}^L + \mathbf{A}\mathbf{m}^D.$$

Finally, rearranging these Equations we obtain the more convenient expressions:

$$(\mathbf{I} - \mathbf{A}')\mathbf{m}^U = \mathbf{m}^L, \tag{2.12}$$

$$(\mathbf{I} - \mathbf{A})\mathbf{m}^D = \mathbf{m}^L. \tag{2.13}$$

The two linear systems defined by Equations 2.12 and 2.13 allow the immediate computation of carbon intensities, taking into account all indirect effects, instead of the recursive formulas 2.4 and 2.7, which would require an infinite number of iterations.

In classical environmental IO analysis, upstream carbon intensities are computed using the row column-matrix product, instead of the more familiar matrix-column vector product, which we use in Eqs 2.12 and 2.13. Converting Eq. 2.12 to row vector format requires transposing the matrix of technical coefficients:

$$(\mathbf{m}^{U})'(\mathbf{I} - \mathbf{A}) = (\mathbf{m}^{L})'$$

Finally, we can obtain the explicit expression usually used to compute total upstream embodied emissions (Miller and Blair 1985; 203):

$$(\mathbf{m}^{U})' = (\mathbf{m}^{L})'(\mathbf{I} - \mathbf{A})^{-1}$$

We shall use the more conventional matrix-column vector multiplication format, however.

Let us refer to $(\mathbf{I} - \mathbf{A}')^{-1}$ and $(\mathbf{I} - \mathbf{A})^{-1}$ as the *Ghosh* and *Leontief inverses* (Ghosh 1958, Leontief and Ford 1970), respectively. Total indirect effects can be disaggregated by expanding these inverses. (An alternative formulation can be found in Miller and Blair (1985, 23).)

The Taylor expansion of function $f(x)$ around y is

$$f_y(x) = \sum_{n=0}^{\infty} \frac{(x-y)^n}{n!} f^{(n)}(y)$$

where $f^{(n)}(x)$ denotes the n-th derivative of $f(x)$. The Taylor expansion of function $f(x) = (1-x)^{-1}$ around 0, since $f^{(n)}(x) = \frac{n!}{(1-x)^{n+1}}$, is given by:

$$f_{y=0}(x) = \sum_{n=0}^{\infty} \frac{x^n}{n!} \frac{n!}{(1-0)^{n+1}} = \sum_{n=0}^{\infty} x^n$$

This expansion is valid if $|x| \ll 1$. Thus, if the norm of a matrix is small, the same reasoning applies and the inverse matrices can be expressed as power series. This condition usually holds in IO matrices, and therefore the

Taylor expansion can be applied:

$$(\mathbf{I} - \mathbf{A}')^{-1} = \sum_{n=0}^{\infty} (\mathbf{A}')^n = \mathbf{I} + \mathbf{A}' + (\mathbf{A}')^2 + (\mathbf{A}')^3 + \dots.$$

$$(\mathbf{I} - \mathbf{A})^{-1} = \sum_{n=0}^{\infty} \mathbf{A}^n = \mathbf{I} + \mathbf{A} + \mathbf{A}^2 + \mathbf{A}^3 + \dots$$

The decomposition of the inverses as power series is conceptually useful since it shows that the total indirect effects are a sum of direct effects (the first term in the series), first-order indirect effects (the second term), second-order indirect effects (the third term) and so on. Plugging the expansion into Eqs. 2.12 and 2.13 we obtain:

$$\mathbf{m}^{\mathbf{U}} = (\mathbf{I} - \mathbf{A}')^{-1} \mathbf{m}^{\mathbf{L}} = \sum_{n=0}^{\infty} (\mathbf{A}')^n \mathbf{m}^{\mathbf{L}} \qquad (2.14)$$

$$\mathbf{m}^{\mathbf{D}} = (\mathbf{I} - \mathbf{A})^{-1} \mathbf{m}^{\mathbf{L}} = \sum_{n=0}^{\infty} (\mathbf{A})^n \mathbf{m}^{\mathbf{L}} \qquad (2.15)$$

Equations 2.14 and 2.15 show that total carbon intensities in turn can be decomposed as a sum of direct carbon intensity and the indirect carbon intensities of increasing order.

The Taylor decomposition is useful if one wishes to identify the source of the indirect effects, in a similar way to what is performed in structural path analysis (Peters and Hertwich 2006). For small matrices, the power series are also useful to compute an explicit approximation of the inverses.

The Taylor decomposition also helps us to prove a result that will be useful later on. Assuming all transaction flows t_{ij} are non-negative, then all the entries in $\mathbf{A}'$ are nonnegative too. Therefore, Eq. 2.14 shows that the upstream carbon intensity of any sector i, m_i^U is a linear combination of the direct emissions of all sectors, m_j^L, with $j = 1, \dots, N$, with nonnegative coefficients.

Since upstream embodied emissions are also linearly dependent on upstream intensity and direct emissions are linearly dependent on direct intensity, both with nonnegative coefficients, it follows that:

$$\frac{\partial e_{ij}^U}{\partial e_k^L} \geq 0, \quad \text{for } i, j, k = 1, \dots, N, \qquad (2.16)$$

if the set of values $\{e_l^L\}_{l=0,\dots,N, l \neq k}$ and $\{t_{ij}\}_{i,j=0,\dots,N}$ is held constant.

In words, Eq. 2.16 states that an increase in the direct emissions of some sectors cannot lead to a decrease in the upstream embodied emissions of a flow, other things held equal. A similar statement could be constructed for downstream embodied emissions.

2.3 Other methods

There are other frameworks besides environmental IO analysis to compute indirect carbon emissions, of which the most important is life cycle assessment (LCA) (Hertwich 2005). LCA, whose guidelines are provided by the ISO standards (ISO 2006), is a process-based technique, which is very similar to IO analysis in its philosophy but differs in two important points.

First, LCA is a bottom-up, while IO is a top-down, approach. With IO analysis all indirect effects are captured, but it is not possible to compute intensities for sectors with more detail than that reported in the source table. In LCA, typically a key product is chosen and the processes 'close' to that product are characterized (where 'distance' is measured along the production chain, as explained in Section 2.2.2) After the system boundaries (i.e. the number of steps upstream in the production chain that will be accounted for) are chosen, the indirect effects are computed using formula 2.4 (Lenzen 2001).

That is, LCA has the advantage over IO analysis in that it allows for an arbitrary zoom in on particular products. However, there is the disadvantage that it requires the arbitrary specification of boundaries and the gathering of that detailed information.

Second, instead of using monetary causality to solve the allocation problem, LCA studies often rely on physical causality. That is, the amount of emissions occurring in a process is distributed among outflows (when computing upstream embodied emissions) in proportion to that outflow's weight, energy content, or some other physical quantity, but not in proportion to its monetary value.

Physical IO analysis (Hoekstra and van den Bergh 2006) is a branch of IO analysis that also follows physical causality (Weisz and Duchin 2006). It requires the use of physical IO tables (which report flows between sectors in physical dimensions), which have been published for some countries but which are not published as frequently as monetary IO tables (Giljum and Hubacek 2004).

Although the mathematical techniques can be similar for these different techniques, the interpretation of the computed quantities is different. LCA takes a 'temporal' definition of a life cycle, where an event occurring upstream is something that took place in the past. IO analysis takes a 'snapshot' definition of a life cycle, where all events occur simultaneously. The difference can be easily understood with an example.

Consider the sale of a car. The LCA definition of its upstream embodied emissions are the emissions associated with the manufacture of that particular car, that might have taken place in a previous year. The IO definition of its upstream embodied emissions includes a fraction of the emissions of the suppliers of the car factory that take place during the current year, and the downstream embodied emissions include the emissions of fuel consumption during the current year and a fraction of the emissions of other emissions that the buyer of the car causes during the current year.

'Hybrid' life cycle analysis consists of the coupling of an IO table, which takes into account higher order indirect effects, and detailed product level data, which takes into account direct and lower order indirect effect. This approach seems to us the most appropriate if one is interested in computing the embodied emissions of particular products, since it combines the flexibility of LCA with the closure of IO analysis.

In the remainder of this book we shall confine ourselves to conventional environmental IO analysis, which we consider appropriate to provide information for embodied emissions of whole regions or of particular sectors. However, we believe the book provides all the tools necessary for the interested reader who wishes to extend the current analysis to the level of products, using hybrid LCA.

3 Carbon indicators

3.1 Review

3.1.1 Direct and indirect emissions

The evaluation of performance of the signatory countries to the Kyoto Protocol is made on the basis of GHG inventories. According to the Kyoto Protocol, a country's GHG inventory includes 'greenhouse gas emissions and removals taking place within national (including administered) territories and offshore areas over which the country has jurisdiction' (Peters and Hertwich 2008b). That is to say, the Kyoto Protocol uses direct emissions as the environmental indicator to establish targets for GHG emissions.

Direct emissions as an environmental indicator, has obvious advantages as it is the most straightforward to compute and is also the most transparent to interpret. However, it also has shortcomings. Some of these concern ambiguities regarding international transportation and transboundary abatement efforts that can be overcome in a satisfactory manner (Peters and Hertwich 2008b).

Even if we assume that such procedural problems are overcome and that all direct emissions can be assigned unambiguously to one and only one economic sector in the world, there persists a larger problem, that of indirect effects. Most carbon emissions are generated to produce goods and services on the one hand and revenue on the other hand. The existence of trade between sectors and between nations creates indirect effects, which were formalized in the previous section, in quantities of embodied carbon emissions.

The goal of micro-level product-oriented environmental policy is often the reduction of total (direct and indirect) carbon emissions embodied in a particular good or service. It is a contradiction that at the macro-level environmental policy of the Kyoto Protocol, countries should only aim to reduce their direct emissions.

Several indicators have been proposed that assign direct and indirect emissions to economic agents, expressing mathematically the qualitative concept of an agent's carbon 'responsibility'. Such indicators pose more problems than direct emissions, as they are computationally more laborious and involve more uncertainty.

However, the most important problem they pose is that there is no such single natural indicator. In the rest of Section 3.1, we shall review in detail the most influential indicators of this type, proposed in the last decade.

The review presented here is focused on carbon indicators, but similar indicators have been proposed for other types of environmental pressure. (Eder and Nadoroslawsky (1999) discuss different types of responsibility principles for environmental pressures in general; Rodrigues and Giljum (2005) characterize material flow indicators; Ebert and Welsch (2004) address environmental indices).

3.1.2 Attributed CO₂ emissions

Kondo *et al.* (1998) criticized the use of territorial or direct GHG emissions in climate change policy, because, on the one hand, there are emissions associated with a country's imports that are not accounted for and, on the other hand, some portions of the emissions in a country are due to satisfying foreign demand.

Kondo *et al.* write that

> [o]ne of the means of reducing the national CO_2 emission, in the conventional sense, is for domestic industries to transfer their production systems outside [the country]. From the global point-of-view, such measures are deplored. Such a measure may lead to the 'carbon leakage' problem, which is to increase the total CO_2 emissions worldwide by the transfer of the production systems to other countries with lower energy-efficiencies for production.
>
> (Kondo *et al.* 1998: 164)

Based on these ideas, these authors propose the indicator of *attributed CO₂ emissions* as the 'national responsible CO_2 emission'.

They note that the national direct CO_2 emissions, E_N, can be decomposed into emissions for domestic demand, A, and for foreign demand, B, $E_N = A + B$. They discriminate, from foreign direct emissions, a component of emissions for domestic demand, C, which, added to domestic emissions for domestic demand, adds up to the CO_2 emissions induced by domestic final demand, $E_D = A + C$.

They consider that a country's direct emissions for domestic demand, A, belong to that country alone but that the other two components, B and C, should be shared between the 'producing' and 'consuming' countries.

That is, they consider fractions p and q, such that $p, q \geq 0$ and $p + q = 1$, and suggest that attributed CO_2 emissions, E_A, be defined as:

$$E_A = A + pB + qC$$

The authors suggest that the allocation parameter p can be adjusted for different commodities, but they do not mention which criteria should be used for that adjustment or whether such criteria should be the same for all countries.

It is illustrative to look at the extreme cases of p. If $p = 0$, then $E_A = E_N$, and the attributed emissions equal national direct emissions. If $p = 1$, then $E_A = E_D$, the attributed emissions equal direct emissions induced by domestic final demand.

The authors comment that

> [e]ven if any industries transfer their production systems to other countries to reduce their domestic CO_2 emissions, we have to add the foreign emissions for imports to the attributed CO_2 emission when we import commodities into [the target country]. So, any industry needs to take counter measures, to reduce CO_2 emissions, such as the introduction or the development of the production systems with lower CO_2 emissions in countries which export commodities, to [the target country]. On the other hand, even if industries produce commodities for export, the associated country should take the responsibility for the resulting emissions.
>
> (Kondo *et al.* 1998: 173)

The calculations presented by the authors are done in the framework of national IO modelling, thus taking into account only domestic indirect effects. Quantities A and B can be unambiguously computed in such a framework, but quantity C is ambiguous.

3.1.3 *CO₂ trade balance*

The work of Munksgaard and Pedersen (2001) was motivated by the situation of Denmark, a small open economy which exports a substantial amount of coal-generated electricity. According to the Kyoto rules for the compilation of GHG inventories, the emissions resulting from the generation of this electricity are assigned to the generating country (Denmark). However, the Danish statistical office reports an alternative GHG inventory in which these emissions, required to produce exports, are not accounted for (instead, they should be accounted for by the importing countries).

Inspired by Proops *et al.'s* (1993) distinction between carbon 'emissions' and carbon 'responsibility', Munksgaard and Pedersen proposed two pure 'producer' and 'consumer' accounting principles, which are in essence similar to the ones proposed by Kondo *et al*. The 'producer' CO_2 account of Munksgaard and Pedersen is the total national direct emissions, and the 'consumer' CO_2 account is the total upstream carbon emissions (TUECES) embodied in final demand.

(The authors' formulas are more complex but they reflect the methodological option of separating domestic and foreign indirect effects of final demand, only to add them up at the end).

The authors subtract the 'producer' from the 'consumer' CO_2 account to obtain a CO_2 trade balance. According to the authors, 'the CO_2 trade balance can be regarded as an estimation of net CO_2 emissions abroad caused by [national] consumption'. They further argue that net CO_2 exporters are disadvantaged, if the establishment of carbon abatement targets in environmental agreements is based on direct emissions only and conclude that 'taking such imbalances into account in foreign trade might reduce the reluctance of some open economies to accept such kinds of agreements' (Munksgaard and Pedersen 2001: 333).

Thus, the authors suggest that a country's allowed emissions in an environmental negotiation should be an intermediate quantity between their direct emissions and their total upstream emissions of consumption. However, the precise value should be subject to negotiation.

3.1.4 Benefit principle and ecological deficit

Ferng (2003) introduces ethical concepts in order to define the quantity of 'responsible anthropogenic CO_2 over-emissions', which should be used to calculate the amount of carbon abatement each country is 'responsible' for.

Thus, according to Ferng, following the principle of 'ecological deficit', it is possible to estimate how much carbon is removed from the atmosphere through natural sinks and to allocate every sink to a particular country.

As Ferng argues, the existing anthropogenic GHG emissions should be allocated to countries according to the 'benefit' principle, and it is the difference between the anthropogenic emissions ('benefit') and natural abatement ('ecological deficit') that defines 'responsible carbon over-emissions'.

We now look with more attention to the 'benefit' principle, which defines the carbon indicator of a region. According to the author '[t]he benefit principle argues that production and consumption activities are both responsible for excessive anthropogenic CO_2 emissions'. The benefit principle is divided into 'consumption' and 'production' benefit.

She proposed as indicator of the CO_2 responsibility of a country the expression $\phi A + (1 - \phi)B$, where $0 \leq \phi \leq 1$ is a weight parameter, A is the 'consumption' responsibility and B is the 'production' responsibility. A is the total upstream embodied emissions of consumption, as was the case in the previously mentioned studies.

The 'production' responsibility, B, instead of being the national direct emissions as previously defined, is now the sum of national direct emissions and of the total upstream embodied emissions of intermediate imports. According to the author,

> [i]n terms of anthropogenic CO_2 emission, a defined country is responsible for: (1) the CO_2 emitted from its domestic production activities no matter where the products are delivered to (emissions occur in the defined country); and (2) the amount of CO_2 emitted in foreign countries when producing the goods and services that are required to deliver the commodities to the production sectors of the defined country.
>
> (Ferng 2003: 128)

This quantity exhibits double counting, meaning that if a country is split into two regions, the sum of the Bs of the two regions will be larger than the B of the whole country, because the emissions embodied in the trade between the regions are accounted for in the former case but not in the latter. These extra emissions were already accounted for in the direct emissions, and thus they are accounted for twice.

3.1.5 Carbon emissions added

Bastianoni *et al.* (2004) also try to present a carbon indicator that takes into account indirect emissions, but they want to escape the consumption–production dichotomy so starkly presented in the previous studies. These authors adopt an approach from embodied energy analysis and propose the carbon emissions added (CEA) indicator.

The authors claim that the CEA indicator is the trade-off between consumption and production accounting principles. In their opinion,

> this approach allows to share the responsibilities among all the interested subjects in an efficacious and fairer way: consumers are taken as responsible for most of the emissions but have the possibility to choose the optimal producer; producers are subject to a minor but precise imputation of responsibility.
>
> (Bastianoni *et al.* 2004: 256)

The CEA indicator is calculated in two steps. First, each country is assigned the total upstream embodied emissions of its imports and its national direct emissions (the 'production' indicator of Ferng [2003]) (i.e., its TUECES). In a second step, the TUECES of each country are normalized, such that the sum of all CEAs equals the total direct emissions of the world.

In the example provided by the authors, there are three countries, A, B and C, with direct emissions of 50, 30 and 20 units, respectively, and there is trade from A to B and from B to C. The total upstream embodied emissions are, respectively, 50, 80 and 100, and the total is 230. Thus, the normalization consists in multiplying the previous quantities by 100/230 to ensure that their total equals the total of direct emissions. Thus, the CEAs of each country are, respectively, 22, 35 and 43 units.

This indicator has no double counting, because it is normalized, but it is not invariant to aggregation. This means that if two countries are joined in a single region, the CEA of the combined region is different from the sum of the CEAs of each country. Let countries B and C be combined as region D. Now the direct emissions of A and D are, respectively, 50 and 50. The total upstream embodied emissions are 50 and 100, and the CEAs are 33 and 67. Since $67 < 35 + 43$, the CEA of the combined region is lower than the sum of the CEAs of the two countries.

3.1.6 Shared responsibility

Gallego and Lenzen (2005) were concerned with the problem of devising an accounting method that allows apportioning of indirect emissions to both producers and consumers while avoiding double counting. Their solution to this problem was to define the quantity of *partial* embodied emissions.

For example, the upstream embodied emissions of the inputs of a sector are added to the direct emissions of that same sector. But now, instead of this whole quantity flowing to the outputs of that sector (as in total upstream embodied emissions), there is a sink in the sector, which retains a fraction of that flow.

The authors equate the amount of partial upstream emissions that are retained by the sectors as producer responsibility and the amount of partial upstream emissions that reach final demand as consumer responsibility. These two quantities sum up to total direct emissions, so there is no double counting.

The authors allowed for an arbitrary fraction of upstream emissions to be retained in each sector (and they also explored the possiblity of accounting for partial downstream emissions).

In another paper, Lenzen *et al.* (2007) proposed a particular rule for the specification of the retained factors: added value's share of total inputs. For example, if the total inputs of a sector are 100 and its added value is 25; and its partial upstream embodied emissions (from inputs plus direct) is 16, then the sector keeps $4 = 16 \times \frac{25}{100}$, and the remaining emissions are distributed among the outputs.

The justification for the added value rule is based on invariance arguments that are incorrect (Rodrigues and Domingos 2008). But the authors also argue that '[v]alue added indicates whether or not a producer has transformed operating inputs in any significant way, and is therefore a good proxy for control and influence over production'.

The authors further state that the shared responsibility has advantages over either full 'consumer responsibility' (= total upstream embodied emissions of consumption) or full 'producer responsibility' (= direct emissions).

> [I]n contrast to full producer responsibility, in shared responsibility, every member of the supply chain is affected by their upstream supplier and affects their downstream recipient, hence it is in all actors' interest to enter into a dialogue about what to do to improve supply chain performance. There is no incentive for such a dialogue in full producer responsibility. In shared responsibility, producers are not alone in addressing the impact (ecological footprint) issue, because their downstream customers play a role too.
>
> (Lenzen *et al.* 2007: 36)

And 'in contrast to full consumer responsibility, shared responsibility provides an incentive for producers and consumers to enter into a dialogue about what to do to improve the profile of consumer products'. (Lenzen *et al.* 2007: 36).

3.1.7 Emissions embodied in trade

Peters and Hertwich (2008a) are interested in how emissions embodied in international trade shape a country's environmental profile.

The authors define *domestic* upstream embodied emissions of an economic flow as the direct and indirect emissions occurring within a country required to generate that flow. The domestic upstream emissions of domestic final demand and exports are identical to the total direct emissions of a country. Since this is true, it is possible to discriminate the domestic emissions embodied in bilateral trade:

$$f_r^d = \sum_s f_{rs} \equiv f_{rr} + \sum_{s, s \neq r} f_{rs}$$

where f_r^d are the direct emissions of country r (which the authors refer to as 'production-based' GHG inventory), and f_{rs} are the domestic emissions embodied in the trade from country r to country s.

The authors define the total emissions embodied in exports (EEE), f_r^e, as:

$$f_r^e = \sum_{s,s \neq r} f_{rs}$$

and, symmetrically, emissions embodied in imports (EEI), f_r^i, as:

$$f_r^i = \sum_{s,s \neq r} f_{sr}$$

Combining these quantities, the authors propose a 'consumption-based' emission inventory defined as 'the total global emissions occurring from economic consumption within a country'. Peters and Hertwich (2008a) claim that such an inventory 'can thus be considered a trade-adjusted version of the production-based inventory. The consumption-based inventory takes the production-based inventory, but deducts the EEE and adds the EEI.' Formally, the consumption-based inventory, f_r^c, is:

$$f_r^c = f_r^d + f_r^i - f_r^e$$

This indicator is based on domestic embodied emissions, and thus only takes into account first-order international indirect effects. For example, let us consider an international linear supply chain involving three countries, A (upstream), B (intermediate) and C (downstream). The consumption-based inventory of country C takes into account the emissions occurring in B, but not those occurring in A. Thus, as in process-based LCA, this indicator truncates the order to which indirect effects are accounted.

This does not happen due to some data deficiency but due to the construction of the indicator proper. This construction has the advantage of not requiring the specification of international inter-industry flows, which are the most problematic component of an MRIO model.

3.2 Comparison

3.2.1 Purpose

A carbon indicator is a number that answers the question 'who has emitted how much of GHG in a given time period?' The indicator can be used to serve a number of other purposes: to allocate permits, to levy taxes, to monitor performance and to direct abatement efforts.

For instance if the direct emissions of a country are x tonCO$_2$, this indicator glosses over the precise location of the emissions but provides substantial information: if y tonCO$_2$ had been allocated to that country, we could compare both numbers to assess its compliance; if a carbon tax is levied, we know how much money that country has to pay; if another country has z tonCO$_2$ emissions, we can compare the productive structure of both countries and decide, on the basis of its costs and benefits, where emission abatement would be more cost-effective.

The indicators we have reviewed here are aimed at different scales: for example, CEA aims at distinguishing the carbon responsibility of different countries along a production chain, while 'shared responsibility' aims at distinguishing the carbon responsibility of firms and final consumers, within a country. However, to some extent, the properties that are required from an indicator are the same for all scales.

On the one hand, there are practical properties that the indicator must satisfy. These are related to how easy it is to measure, i.e., whether it requires substantial data and has high uncertainty; and how easy it is to define in the first place, i.e., whether it is ambiguous or clearly defined. On the other hand, there are ethical issues to be considered. Indicators of indirect emissions are often referred to as 'carbon responsibility', which is an expression that by itself contains a value judgement. All indicators of indirect emissions are constructed by assigning some form of emissions embodied in flows to economic agents, and the choice of which type of embodied emissions (e.g., total, partial or domestic, upstream or downstream) is assigned to whom is essentially ethical.

We shall compare the indicators reviewed above according to three particular properties: measurement issues; scale invariance; and accounting of indirect effects.

3.2.2 Measurement

The problem of measurement concerns the amount of information required and its associated uncertainty. Most indicators claim to take into account total upstream embodied emissions (even though combined in different ways). The proper accounting of total embodied emissions requires a full MRIO model, which is the type of model that involves the largest amount of data and uncertainty; (although substantial simplifications are common, such as using only national tables).

There were two indicators that were based on other types of embodied emissions: partial embodied emissions (in the case of 'shared responsibility') and domestic embodied emissions (in the case of the 'consumption-based carbon inventory'). The latter requires the same data as total embodied emissions but involves yet more uncertainty in the calculations. The former requires somewhat less data and uncertainty, as it does not require a full MRIO but only domestic tables and exports discriminated by region (but not sector).

Thus, on the issue of measurement, 'consumption-based carbon inventory' stands out, while all the other indicators are essentially equivalent.

3.2.3 Scale invariance

We live in a hierarchical world: countries are sets of regions, which are sets of municipalities, etc.; economic sectors are sets of firms, which are sets of departments, etc. Thus, a practical property that an indicator must satisfy is scale invariance, the property that if one agent is disaggregated into several agents, the sum of the indicators of the composing agents matches the indicator of the aggregated agent.

If this property is not verified, then the indicator is scale-dependent, and it is not possible to compare, for example, the performance of a country with that of a region (either from the same or other country). This is a disadvantage because this way it is not possible to determine so as best practice, to replicate it in another region or country.

The property of scale invariance is not verified by some indicators ('benefit principle and ecological deficit', 'attributed CO_2 emissions' and CEA). The remaining indicators satisfy this property.

3.2.4 Total vs partial embodiment

Most indicators are based on total embodied emissions and thus are based on the ethical principle that an agent is responsible for the carbon emissions occurring throughout the life cycle of some particular good or service (usually, but not always, of final demand).

The accounting of total embodied emissions originates a very clear suggestion for policy action: by choosing economic inputs, on the basis of their total carbon intensity, the economic agent is manipulating its own carbon indicator and it can be sure of the 'true' impact of its actions concerning direct emissions. Let us consider that the carbon indicator is the total upstream embodied emissions of final consumption. Let us consider now that an agent has the option of choosing product A, whose TUECEF is 20 tonCO$_2$, and product B whose TUECEF is 10 tonCO$_2$. In a static perspective, the agent knows that by choosing B rather than A it is saving the world 10 tonCO$_2$. (In a dynamic perspective things are more complex, of course.)

Two of the indicators presented use partial embodied emissions. 'Shared responsibility' is based on upstream partial embodied emissions, in which there is a sink of embodied emissions in every sector (with a sector-specific factor). 'Consumption-based carbon inventory' is based on domestic embodied emissions, in which, considering a specific country A, the emissions occurring inside A are assigned to its outputs, using the expression of total upstream embodied emissions (that is, without a sink). Since the emissions embodied in the imports into A are not accounted for, this is also a type of partial embodied emissions (domestic embodied emissions).

The accounting of partial embodied emissions does not guarantee that the 'true' impact of an agent's actions and his carbon indicator move in the same direction. The emissions occurring far upstream in a value chain are not reflected in the partial embodied emissions, and thus the choice of a final product can lead to a reduction in the value of the indicator, and simultaneously, to an increase in real emissions.

Let us consider the following example concerning domestic embodied emissions (a similar example can be constructed for partial embodied emissions). Let product A result from a 2-country production chain, X and Y, with 10 $tonCO_2$ of direct emissions occurring in either country. The domestic embodied emissions of A are 10 $tonCO_2$, as only emissions occurring in country Y are accounted for. Let us now consider product B, from a 2-country production chain, W and Z, with respective direct emissions of 0 and 20 $tonCO_2$. The domestic embodied emissions of B are now 20 $tonCO_2$. By choosing A rather than B, a third country can improve its performance by using an indicator based on domestic embodied emissions, but the real emissions resulting from either product A and B are identical.

3.2.5 Conclusions

From the indicators reviewed, we consider that none of these offers simultaneously the properties of scale invariance and total accounting of indirect effects. There are other properties, which we consider important, but the indicators reviewed so far are identical, in regard to those properties.

A particular point that unifies all the reviewed indicators is that they were proposed on the basis of a verbal statement of their desired properties, but not of a formal derivation. That is, the authors mention properties they consider the indicator should possess, and present the actual formulation of the indicator, but they neither guarantee that the indicator actually possesses those desired properties nor prevents some other indicator from possessing them.

In the following chapter we shall present another carbon indicator, proposing the pursuit of a different strategy.

4 Carbon responsibility

4.1 Properties

4.1.1 Axiomatic approach

In this chapter we derive the indicator of carbon responsibility, following the work reported in Rodrigues *et al.* (2006). First, we shall propose a number of properties that we consider that a good indicator of indirect carbon emissions should possess, and afterwards we will derive the mathematical formulation of this indicator, which we call *carbon responsibility*.

Ideally, society at large would need to agree with the properties of carbon responsibility, in order to accept the resulting indicator. However, if different economic agents were allowed to decide independently the properties of the indicator, aware of their own situation and interests, a consensus would probably be imposssible to achieve, since people in different situations would probably have contradictory interests. Thus, we follow John Rawls's idea of the 'veil of ignorance' (Rawls 1971), in order to establish the properties of carbon responsibility.

We consider that carbon responsibility has the properties that would be chosen if, in an 'original position', society had to agree upon them in order to implement environmental policy at a later stage, and every individual involved in the decision process was behind the 'veil of ignorance', being unaware of what his own carbon responsibility would be, compared to that of others.

This is a thought experiment only: we are not in an 'original position', and every individual has specific economic and environmental characteristics that allow him/her to predict the carbon responsibility that would result from his/her choice of properties. That is, the 'veil of ignorance' cannot be enforced in reality. Thus, the hypothetical situation of an 'original position' behind the 'veil of ignorance' was an inspiration for the specific properties proposed here, but these were conceived by the actual authors of this book and result from several discussions with several people, yet by no means 'society' at large.

We now introduce the notation and nomenclature to be used in the rest of the chapter. For parsimony of language, we shall consider that the *world* is divided in a set of non-overlapping *regions* (where region could be any economic agent). Each region is in turn divided in a set of non-overlapping sectors. W is the world and R_W is the set of all regions in the world. S_W is the set of all internal sectors (i.e., excluding sector 0) and S_k is the set of internal sectors of country k. F_k is the set of flows (i.e., transactions between sectors) in which country k is involved, i.e., the set of flows (ij) such that either i or j belong to S_k, and F_W is the set of all flows in the world.

All direct carbon emissions are assigned to some internal sector i, e_i^L and E_k^L are the direct (or local) emissions of country k, defined as the sum of the emissions of all sectors that compose it:

$$E_k^L = \sum_{i \in S_k} e_i^L$$

It follows logically that the total of direct emissions in the world is E_W^L. We shall use E_k to denote the *total carbon responsibility* of country k, which we shall construct as a function of two auxiliary indicators, E_k^C and E_k^P, respectively the *consumer* and *producer carbon responsibility* of country k. The consumer and producer responsibilities are in turn constructed on the basis of upstream and downstream embodied emissions, respectively, E_k, E_k^C and E_k^P. The work reported in this chapter differs from Rodrigues *et al.* (2006) because here the two auxiliary indicators are derived prior to total carbon responsibility, while in Rodrigues *et al.* (2006) they were derived afterwards.

In the following subsection we list the desired properties of carbon responsibility.

4.1.2 Consistency requirements

We consider there are three properties that can be understood as consistency requirements, and therefore should not be contentious: scale invariance, normalization and monotonicity.

The requirement for scale invariance stems from the fact that the world we live in has the topology of a nested hierarchy, crossing different scales: a country is a collection of regions, regions are collections of municipalities and so forth.

The implementation of actual policies requires that the carbon indicator is not affected by the political landscape: the carbon responsibility of a country or the carbon responsibility of the set of its regions should be the same. Otherwise, it would be possible to alter the carbon performance of a country (as displayed by its carbon responsibility) merely by the shifting of borders, without any real improvement. Indicators that are not scale-invariant, like Total Material Requirements or TMR (Rodrigues and Giljum 2005), display such shifts when there are secessions or annexations (e.g., Yugoslavia or Germany in the early 1990s).

Property 1 (Scale invariance)

If k, k' and k'' are countries, such that $S_k = S_{k'} \cup S_{k''}$ then $E_k = E_{k'} + E_{k''}$, $E_k^C = E_{k'}^C + E_{k''}^C$ and $E_k^P = E_{k'}^P + E_{k''}^P$

Or, in words, if country k is the union of countries k' and k'', then the carbon responsibility of k must equal the carbon responsibilities of k' and k''. Property 1 demands that the carbon responsibility of a country is the sum of the carbon responsibilities of the regions that compose it; likewise for consumer and producer responsibility. In plain language, scale invariance requires that the total equals the sum of the parts.

The second consistency requirement we impose is normalization.

Property 2 (Normalization)

The total, consumer and producer carbon reponsibility of the world are all identical to the total direct carbon emissions of the world, $E_W = E_W^C = E_W^P = E_W^L$.

This property requires that total, consumer and producer carbon responsibility all have the same physical dimensions as direct emissions. This property is conceptually helpful, because it transforms the problem of defining carbon responsibility into a 'cake eating' problem, i.e., allocating a fraction of the direct emissions of the world to each particular country.

This is a weak property and hard to object to since, even if the indicator is originally constructed in such a way that it does not satisfy Property 2, is is always possible to construct a related indicator that does satisfy that property, by multiplying all carbon responsibilities by a common factor (equal to $E_W^L / \sum_{k \in R_W} E_k$).

The final consistency requirement we impose is, in a sense, the definition proper of a carbon indicator. A carbon indicator displays the carbon performance of an economy: if it goes up, direct emissions should be going up somewhere and vice versa.

Property 3 (Monotonicity)

The total, consumer and producer carbon responsibilities of country k, E_k, E_k^C and E_k^P, are, respectively, unspecified functions

$$E_k = g(E_k^C, E_k^P)$$

$$E_k^C = g^C \left(\{e_i^L\}_{i \in S_W}, \{t_{(ij)}\}_{(ij) \in F_W} \right)$$

and

$$E_k^P = g^P \left(\{e_i^L\}_{i \in S_W}, \{t_{(ij)}\}_{(ij) \in F_W} \right)$$

such that

$$\frac{\partial g}{\partial E_k^C} \geq 0, \ \frac{\partial g}{\partial E_k^P} \geq 0, \ \frac{\partial g^C}{\partial e_i^L} \geq 0, \ \frac{\partial g^P}{\partial e_i^L} \geq 0, \quad for \ i \in S_W$$

Or, in words, the consumer and producer carbon responsibilities of any country cannot go down if the direct emissions of *any* given sector in the world go up, while the direct emissions of all other sectors remain constant and the economic structure remains the same. The total carbon responsibility of a country cannot go down if either the consumer or producer responsibility of that same country goes up, and the other argument remains constant.

Note that:

$$\frac{\partial E_k}{\partial e_i^L} = \frac{\partial g}{\partial E_k^C} \frac{\partial g^C}{\partial e_i^L} + \frac{\partial g}{\partial E_k^C} \frac{\partial g^C}{\partial e_i^L} \geq 0, \quad for \ i \in S_W \quad and \quad k \subset W$$

Monotonicity does *not* require that the indicator to be a function of *all* direct emissions. If the indicator is independent of some particular flow, Property 3 is still satisfied.

Property 3 also demands that the functional forms of functions g, g^C and g^P are independent of the labelling of regions (otherwise each region k could have a particular g_k, for example). This means that the functional form of the indicator is the same for all regions.

4.1.3 Ethical considerations

The practical properties reported so far do not suffice to define a single indicator of carbon responsibility. So we need to introduce the following properties, which, while being formal (i.e., mathematical), express ethical considerations.

The first of such properties is the accounting of total indirect effects. By this property, we mean that the indicator must be constructed using total embodied carbon emissions (whether upstream or downstream), and not some form of partial embodied emissions.

The reason for the introduction of consumer and producer carbon responsibilities becomes apparent now, since these are indicators built using either upstream or downstream embodied emissions.

Property 4 (Total indirect effects)

The consumer carbon responsibility of country k, E_k^C, is an unspecified function

$$E_k^C = f^U\left(\{e_{ij}^U\}_{(ij)\in F_k^U}\right)$$

and the producer carbon responsibility of country k, E_k^P, is an unspecified function

$$E_k^P = f^D\left(\{e_{ij}^D\}_{(ij)\in F_k^D}\right)$$

where e_{ij}^U and e_{ij}^D are given by 2.4 and 2.7, and F_k^U and F_k^D are unspecified sets of flows such that $F_k^U, F_k^D \subseteq F_k$

In the case of open economies, countries' actions are always interdependent. In particular, a country often benefits from the carbon emissions taking place in another country (e.g., by importing carbon intensive goods). Property 4 requires that both direct and total indirect carbon emissions must be taken into account in the construction of carbon responsibility, but it does not specify how. That is, this property neither specifies the functional form of f^U and f^D nor the sets F_k^U and F_k^D.

However, the embodied emissions that can be considered are only those embedded in the flows in which the country is involved. That is, $F_k^U, F_k^D \subseteq F_k$, where F_k is the set of flows (ij) where either i or j belong to S_k. This can be understood as a principle of 'locality', since the responsibility of each country only depends on 'local' TUECEFs or TDECEFs, i.e., emissions embodied in flows in which one of that country's sectors is involved.

The choice of inputs and outputs (i.e., from whom to buy and to whom to sell) is a viable option for carbon policy. If the indicator does not take total indirect effects into account, it is prone to manipulation, meaning that a country can take an option that decreases its carbon responsibility while leading to higher total emissions.

If the carbon indicator satisfies accounting of total indirect effects, it will not display 'carbon leakage'. Carbon leakage occurs when a polluting industry is moved abroad and its carbon intensive goods are then imported. If the indicator is a function of the economic flows at the end of this supply chain (e.g., the final consumption goods), and if there is no change in the supply chain except the geographical relocation of the factory, the carbon indicator does not change.

Note that Property 4 also imposes that f^U and f^D are independent of the region, and thus their functional form is identical for all regions. Note also that Properties 4 and 3 are not inconsistent: functions f^U and f^D take e_{ij}^U's and e_{ij}^D's as arguments, but according to Eqs 2.4 and 2.7, these quantities are in turn functions of e_i^L's and t_{ij}'s, as imposed by Property 3 upon g^C and g^P.

The next property we consider is economic causality. Economic causality is a complement of Property 4, because Equations 2.4 and 2.7 do not uniquely characterize all the embodied emissions TUECEFs and TDECEFs – as we have seen in Chapter 2.

Property 5 (Economic causality)

All economic flows leaving sector i, where $i \in S_W$, have the same upstream carbon intensity, m_i^U, and all economic flows entering sector i have the same downstream carbon intensity, m_i^D, as defined in Section 2.2.2 and given by Eqs. 2.6 and 2.10.

The intuition for Property 5 is that, in all relevant situations, carbon emissions occur as a by-product of economic activities, i.e., the production or consumption of goods and services. Thus, the economic share of an output (resp. input) among all outputs (resp. inputs) is an acceptable proxy for the share of carbon responsibility of that output (resp. input) among all outputs (resp. inputs).

This assumption has the advantage of being consistent with the standard IO computation methods and being easy to extend to other situations, since monetary data is usually available. However, this assumption is at heart an ethical one, and it can be replaced by others without altering the theoretical results that follow.

Physical causality is a valid alternative to economic causality, in which the embodied emissions are shared by all flows leaving or entering a sector according to that flow's weight share of all outputs or inputs. For practical purposes this would require using a Physical Input–Output Table or PIOT for the calculations, as discussed in Chapter 2.

Property 6 (Symmetry)

If the consumer and producer responsibilities are interchanged, total carbon responsibility remains unchanged. That is, for country k,

$$E_k = g\left(E_k^C, E_k^P\right) = g\left(E_k^P, E_k^C\right)$$

In words, Property 6 states that g must remain invariant if its arguments are interchanged. Therefore, the functional form of g is symmetrical.

We require this property, because we did not find any justification to favour either upstream or downstream arguments, which form the consumer and producer carbon responsibilities. All of the indicators reported so far take a strictly upstream perspective, but we do not find any strong reason to take into account upstream indirect effects but not downstream ones.

Upstream indirect effects manifest themselves in the choices of economic inputs and are accumulated in final consumption. Likewise, downstream indirect effects manifest in the choice of economic outputs and are accumulated in primary inputs. We believe that, from an ethical perspective, it is equally legitimate to demand someone to take responsibility for the upstream carbon emissions of its consumption as it is legitimate to ask that person to take responsibility for the downstream carbon emissions of his/her revenues.

Some people criticized this property, because they misunderstood it as a claim about the structure of real economies. Still other people argue that output choices (and thus downstream embodied emissions) are more constrained than input choices (and thus upstream embodied emissions), and thus that symmetry should be refused.

If the reader is not interested in such a unique indicator of carbon responsibility, symmetry is not necessary. However, if the reader is interested in a unique indicator, we believe symmetry is the correct property to define it.

4.2 Derivation

4.2.1 Preliminaries

In this section we derive the indicator of carbon responsibility of a generic country k, E_k, that verifies the six properties defined in Section 4.1.

We begin by considering E_k a totally free and undefined function, and invoke properties to narrow the scope of possible functional forms and arguments, until E_k is uniquely defined.

The derivation involves only elementary calculus and arithmetic operations. The only technique that might be unfamiliar to the reader is the use of differentials, which is a technique borrowed from thermodynamics, whose rationale we briefly outline here. If f is a continuous differentiable function taking independent arguments x and y, $z = f(x,y)$, we can study the effect that a small variation in the arguments causes on the result, using the rule of partial derivation:

$$dz = \frac{\partial f}{\partial x} dx + \frac{\partial f}{\partial y} dy$$

where dx, dy and dz are small. In this case the differentials dx, dy and dz are dependent, meaning that for any particular values of dx and dy, the previous expression imposes the value of dz. Now consider that we arbitrarily set $dz = 0$, thus restricting movement in the 3-D manifold to horizontal movements only. The previous expression can be cast as:

$$0 = Adx + Bdy$$

The remaining differentials, dx, dy are independent, meaning that dx can change while dy remains constant and vice versa, and thus, for movement to remain horizontal, irrespective of dx and dy it is necessary that $A = B = 0$. In order for such an expression to be true in general, i.e., for several independent variables to be independent and their weighted sum to be 0, it is necessary that all weighting factors be 0 too. If the weighting factors are not 0, the expression can only hold if dx and dy are dependent, which is a contradiction.

Our strategy is to obtain expressions in which weighted sums of independent differentials are set to 0, to obtain constraints on the weighting factors.

For convenience of notation, let

$$E_k^U = \sum_{i \in S_k} e_{i0}^U \text{ and } E_k^D = \sum_{i \in S_k} e_{0i}^D \tag{4.1}$$

denote, respectively, the total upstream embodied emissions of final consumption and the total downstream embodied emissions of primary inputs of country k. (Following the convention of Chapter 2, 0 denotes the external sector.) As we saw in Chapter 2, these quantities have the property that, when applied to the whole world, they match total direct emissions,

$$E_W^U = E_W^D = E_W^L \tag{4.2}$$

We can combine Eqs. 4.1 and 4.2 with Property 2 to obtain the following expressions that will later prove convenient:

$$E_W^U = E_W^C \tag{4.3}$$

$$E_W^D = E_W^P$$

We shall also make use of Eqs. 2.4 and 2.7, and 2.6 and 2.10, which are invoked in Properties 4 and 5.

4.2.2 Responsibility of consumption and production

We can combine Property 1 (scale invariance) with Eq. 4.3 to obtain:

$$E_W^U = \sum_{k \subset W} E_k^C \tag{4.4}$$

$$E_W^D = \sum_{k \subset W} E_k^P$$

These equations mean that the upstream embodied emissions of final demand equal the sum of the consumer responsibilities of all regions, and the downstream embodied emissions of primary inputs equal the sum of the producer responsibilities of all regions. Expressions 4.4 can in turn be combined with Property 4 (total accounting of indirect effects) and Eq. 4.1 to obtain:

$$\sum_{i \in S_W} e_{i0}^U = \sum_{k \subset W} f^U(\{e_{ij}^U\}_{(ij) \in F_k^U}) \tag{4.5}$$

$$\sum_{i \in S_W} e_{0i}^D = \sum_{k \subset W} f^D(\{e_{ij}^D\}_{(ij) \in F_k^D})$$

The sets F_k^U and F_k^D are unspecified subsets of F_k (Property 4). F_k can be decomposed into five subsets: domestic flows, imports (flows from other countries), exports (flows to other countries), flows to final demand and flows from primary inputs. We now differentiate the first line of Eq. 4.5, which relates consumer carbon responsibility and upstream embodied emissions, isolating these several subsets.

$$
\sum_{i \in S_W} de_{i0}^U = \sum_{k \subset W} \left(\sum_{i,j \in S_k} \left(\frac{\partial f^U}{\partial e_{ij}^U} de_{ij}^U \right) + \sum_{i \in S_k} \left(\frac{\partial f^U}{\partial e_{i0}^U} de_{i0}^U \right) \right.
$$

$$
\left. + \sum_{i \in S_k, j \in S_{k'}, k' \neq k} \left(\frac{\partial f^U}{\partial e_{ij}^U} de_{ij}^U \right) + \sum_{i \in S_k, j \in S_{k'}, k' \neq k} \left(\frac{\partial f^U}{\partial e_{ij}^U} de_{ij}^U \right) \right) \tag{4.6}
$$

Note that the upstream embodied emissions of flows from primary inputs are zero and thus the right hand side of Eq. 4.6 has only four elements. Eq. 4.6 can be rearranged as:

$$
0 = \sum_{k \subset W} \left(\sum_{i,j \in S_k} \left(\frac{\partial f^U}{\partial e_{ij}^U} de_{ij}^U \right) \right)
$$

$$
+ \sum_{k, k' \subset W, k' \neq k} \left(\sum_{i \in S_k, j \in S_{k'}} \left(\frac{\partial f^U}{\partial e_{ij}^U} + \frac{\partial f^U}{\partial e_{ij}^U} \right) de_{ij}^U \right) \tag{4.7}
$$

$$
+ \sum_{k \subset W} \left(\sum_{i \in S_k} \left(\frac{\partial f^U}{\partial e_{i0}^U} - 1 \right) de_{i0}^U \right)
$$

The first line of Eq. 4.7 contains the differentials of domestic inter-industry trade flows, the second line contains the differentials of international inter-industry trade flows, and the third line contains the differentials of flows to final demand.

In Eq. 4.7 there are $S_W(S_W + 1)$ differentials (S_W^2 of them correspond to e_{ij}^U, where $i, j \in S_W$ and S_W correspond e_{i0}^U, where $i \in S_W$). These differentials are independent for the following reason. According to Properties 4 and 5, the e_{ij}^U are defined on the basis of $S_W(S_W + 1)$ independent t_{ij} and S_W e_i^L., which means that the number of source variables (t_{ij} and e_i^L) is larger than the number of derived variables e_{ij}^U. As such, the variations de_{ij}^U are independent.

Therefore, for Eq. 4.7 to hold, i.e., for the left hand side to be zero, the coefficients of all differentials must be zero. This leads to a contraction of the set F_k^U and a clarification of the functional form of f_k^U.

More precisely:

$$\frac{\partial f^U}{\partial e_{ij}^U} = 0, \qquad\qquad i, j \in S_k;$$

$$\frac{\partial E_k^C}{\partial e_{ij}^U} = -\frac{\partial E_{k'}^C}{\partial e_{ij}^U}, \qquad\qquad i \in S_k, j \in S_{k'}, k' \neq k; \qquad (4.8)$$

$$\frac{\partial f_k^U}{\partial e_{i0}^U} = 1, \qquad\qquad i \in S_k$$

In the second line of Eq. 4.8, f^U was replaced by E_k^C and $E_{k'}^C$ to avoid ambiguities.

The first line of Eq. 4.8 tells us that flows (ij), where $i, j \in S_k$, cannot be part of F_k^U. That is, the consumption responsibility of a region is independent of the transfer of upstream embodied emissions among its sectors.

The second line of Eq. 4.8 tells us that the derivatives of E_k^U with regard to the upstream emissions embodied in international trade flows must be crossed. This implies that if such flows are arguments of f^U then:

$$\frac{\partial E_k^C}{\partial e_{ij}^U} \leq 0 \qquad (4.9)$$

for some flow $(ij) \in F_W$.

However, according to Property 3 (monotonicity),

$$\frac{\partial E_k^C}{\partial e_l^L} \geq 0 \qquad (4.10)$$

for sector l, if the values in the sets $\{e_i^L\}_{(i) \in S_W, i \neq l}$ and $\{t_{ij}\}_{(ij) \in F_W}$ are held constant.

According to Properties 4 and 5, through Eqs. 2.3–2.6, we can invoke Eq. 2.16:

$$\frac{\partial e_{ij}^U}{\partial e_l^L} \geq 0 \qquad (4.11)$$

for $i, j, l \in S_W$.

Take, without loss of generality, one of the e_{ij}^U for which the derivative in Eq. 4.9 is strictly negative (non-zero). This flow is a function of a certain set (established by the set of economic flows t_{ij}) of local emissions e_l^L. From Eq. 4.11, we know that e_{ij}^U is an increasing function of each of these e_l^L. Suppose, again without loss of generality, that we increase the value of one of these e_l^L. The value of e_{ij}^U will increase and, according to Eq. 4.9, E_k^C will decrease, hence violating monotonicity.

So, we have shown that the derivative in Eq. 4.9 can only be zero. Therefore, flows (ij), where $i \in S_k$ and $j \in S_{k'}$ with $k' \neq k$, cannot be part of F_k^U.

The last line of Eq. 4.8 tells us that flows $(i0)$, where $i \in S_k$, are admissible elements of F^U. Integration of that expression implies that

$$f^U = e_{i0}^U + K_i, \text{ for } i \in S_k$$

where K_i is a function of all e_{j0}^U, with $j \neq i$. Equalling all these expressions for f^U we conclude that:

$$f^U = \sum_{i \in S_k} e_{i0}^U + K = E_k^U + K,$$

where K is a constant. Property 1 imposes that $E_k^C = E_{k'}^C + E_{k''}^C$, and according to definition 4.1 $E_k^U = E_{k'}^U + E_{k''}^U$. Thus, if function f^U is to be scale invariant, it is necessary that $K = 2K$, and so

$$K = 0$$

Thus, we have found that:

$$E_k^C = E_k^U \tag{4.12}$$

A similar reasoning, starting with Eq. 4.5 but now deriving the second line, which relates producer carbon responsibility and downstream embodied emissions, would show that:

$$E_k^P = E_k^D \tag{4.13}$$

That is, the consumer responsibility of a region is the sum of the upstream embodied emissions of its final consumption, and the producer responsibility of a region is the sum of the downstream embodied emissions of its primary inputs.

4.2.3 Total responsibility

So far we have used all properties except Property 6 (symmetry) and were able to define the auxiliary indicators of consumer and producer carbon responsibility. We shall now combine these indicators to obtain the indicator of total carbon responsibility.

Property 2 implies that $E_W = KE_W^U + (1-K)E_W^D$, where K is a real number. From Property 1 we know that $E_W = \sum_{k \subset W} E_k$. From Property 3 we also know that $E_k = g(E_k^C, E_k^P)$. Combining all this information with Eqs. 4.12 and 4.13 we can write:

$$KE_W^C + (1-K)E_W^P = \sum_{k \subset W} g(E_k^C, E_k^P)$$

The previous expression can be rearranged as:

$$0 = \sum_{k \subset W} \left(g(E_k^C, E_k^P) - KE_k^C - (1-K)E_k^P \right) \tag{4.14}$$

If we differentiate Eq. 4.14 we obtain:

$$0 = \sum_{k \subset W} \left(\frac{\partial g}{\partial E_k^C} dE_k^C + \frac{\partial g}{\partial E_k^P} dE_k^P - KdE_k^C - (1-K)dE_k^P \right) \tag{4.15}$$

$$= \sum_{k \subset W} \left(\left(\frac{\partial g}{\partial E_k^C} - K \right) dE_k^C + \left(\frac{\partial g}{\partial E_k^P} - (1-K) \right) dE_k^P \right)$$

Equation 4.15 has a total of $2S_W$ differentials, which are dependent on a total of $S_W(S_W + 2)$ independent source variables, and are thus independent. Thus, for Eq. 4.15 to hold it is necessary that:

$$\frac{\partial g}{\partial E_k^C} = K$$

$$\frac{\partial g}{\partial E_k^P} = 1 - K$$

Integrating the previous expressions we find that:

$$E_k = KE_k^C + (1-K)E_k^P + K'$$

where K' is the integration constant.

Application of Property 1 along similar lines of Section 4.2.2 shows that

$$K' = 0$$

Finally, Property 6 (symmetry) imposes that

$$E_k = KE_k^C + (1-K)E_k^P = (1-K)E_k^C + KE_k^P,$$

which implies that $K = 1 - K$ and therefore:

$$K = \frac{1}{2}.$$

Thus, we find that the total carbon respnsibility of a region is the arithmetic mean of its consumer and producer carbon responsibilities:

$$E_k = \frac{1}{2}E_k^C + \frac{1}{2}E_k^P \tag{4.16}$$

Equation 4.16 is the key result of this chapter.

4.2.4 Results

Total carbon responsibility of a country is an indicator that satisfies the six properties defined in Section 4.1, and its mathematical formulation is given by Eq. 4.16. The total carbon responsibility of country k, E_k, is the arithmetic mean of two quantities:

(1) Consumer carbon responsibility of country k, E_k^U, which is the sum of the total upstream emissions embodied in the final demand of that country (i.e., the sum of all e_{i0}^U, where $i \in S_k$);
(2) Producer carbon responsibility of country k, E_k^D, which is the sum of the total downstream emissions embodied in the primary inputs of that country (i.e., the sum of all e_{0i}^D, where $i \in S_k$).

5　Multi-regional IO model

5.1　Environmental MRIO models

We use a multi-regional input–output (MRIO) model to compute embodied carbon emissions. Several environmental global MRIO models have previously been constructed, as reviewed by Wiedmann *et al.* (2007).

Ahmad and Wyckoff (2003) and Giljum *et al.* (2008) report MRIO models based on the OECD database (38 regions, 48 sectors).

Lenzen *et al.* (2004), Peters and Hertwich (2006), Weber and Matthews (2007) and Wiedmann *et al.* (2008b) report MRIO models developed from national data sources, with a small number of regions (fewer than ten) and a large number of sectors (over 100 for at least some regions).

Other papers (Ackerman *et al.* 2007, Munksgaard and Pedersen 2001, Peters and Hertwich 2008a, Proops *et al.* 1999) report embodied emissions wihin the (IO) framework, but not using a full MRIO model, and thus not capturing all indirect effects.

(Other models have been developed to compute embodied emissions outside the IO framework, such as Computable General Equilibrium or CGE models [Kainuma *et al.* (2000)]; linear programming [Duchin (2005)]; or simulation models [Lutz *et al.* (2005)].)

We report a full environmental MRIO model based on the GTAP database (87 regions, 57 sectors), as in the recently published papers by Hertwich and Peters (2009) and Andrew *et al.* (2009). Our work differs from that reported in those papers in three main aspects: we report downstream embodied emissions; we use a different method to build the MRIO from the source data; we make an uncertainty analysis of the estimation of international inter-industry flows.

5.2 Data

5.2.1 Sources

The data that we need to compute embodied emissions forms the multi-regional IO (MRIO) model of the world. This model consists (*i*) in a partitioning of the world into a set of regions (either countries or groups of countries), which in turn are partitioned into a set of sectors and (*ii*) in the quantification of all monetary transactions between sectors, t_{ij}'s and the direct carbon emissions occurring in each sector, e_i^L's.

We take our source data from GTAP (more information available at http://www.gtap.agecon.purdue.edu), a research network that gathers national IO data and international trade data to produce a global database, primarily used in a global computable general equilibrium model. We use the GTAP 6 database (Dimaranan 2006) to build the MRIO model, which reports monetary and direct emissions data for the year 2001. (Brockmeier [2001] and McDonald and Thierfelder [2004] were helpful in clarifying the GTAP structure.)

The GTAP direct emissions data (Lee 2007) reports CO_2 emissions resulting from fuel consumption, detailed by firm or household sector, for all regions. A comparison of regional total emissions with the UNFCCC report (UNFCCC 2005) shows major inconsistencies in some regions. Thus, we used the emissions data processed by Glen Peters, corrected on the basis of national emissions reports and previously used in Peters and Hertwich (2008a). This data defines the vector of direct emissions $\mathbf{e}^L$. The GTAP monetary data is more complex and is described in the remainder of this subsection.

GTAP reports transactions for $N_R = 87$ world *regions* (countries and aggregations of countries). For each region GTAP reports transactions for $N_S = 57$ industry (or firm) sectors, $N_F = 3$ *final demand* sectors (investment, government and private households) and $N_E = 5$ *primary input* or endowment sectors (land, unskilled labour, skilled labour, capital and natural resources).

The GTAP data implicitly considers three additional sectors per region: the *tax*, *subsidy* and *regional household* sectors. The tax and subsidy sectors are required for balancing purposes, to explain otherwise unaccounted inputs or outputs. The regional household collects all primary inputs and taxes, and also pays all final demand and subsidies. Figure 5.1 tries to clarify this structure.

Thus, the total number of internal sectors considered is $N_R \times (N_S + N_F + N_E + 2)$ (2 accounts for the tax and subsidy sectors), which is the length of $\mathbf{T}$, the (square) matrix of inter-sectoral transactions, of $\mathbf{e}^L$, the vector of

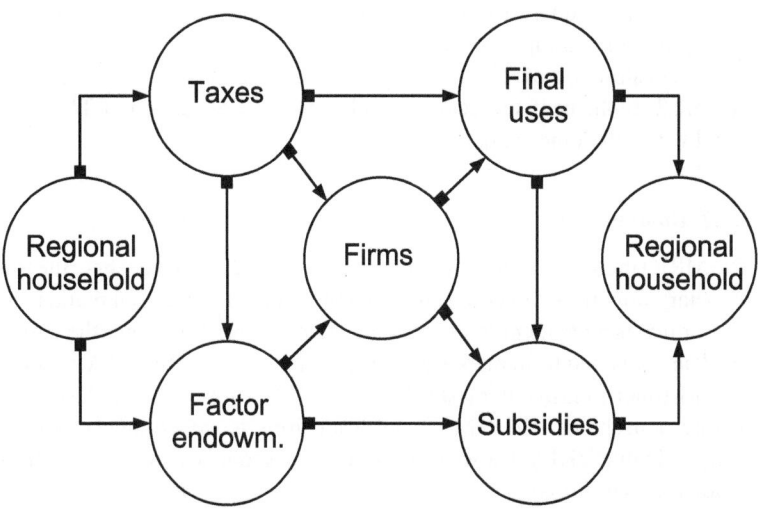

Figure 5.1 Structure of transactions in the GTAP model. ▲ represents the direction of goods and services, and ■ represents the direction of payments.

direct emissions, of **x**, the vector of total output, of **y**, the vector of 'final outputs' and of **v**, the vector of 'primary inputs'. The external sector, 0, is the 'regional households' sector that receives the $v_i = t_{0i}$ and delivers the $y_i = t_{i0}$ flows.

Thus, matrix **T** not only accounts for inter-industry transactions but also for transactions between all internal sectors including final demand sectors, endowment sectors and the tax and subsidy sectors. The area of **T** accounting for inter-firm transactions is dense, while the area accounting for transactions between firms and the other types of sectors is sparse (meaning that it has many zeros).

GTAP reports data in three types of prices: market (or sellers') prices, agents' (or buyers') prices and world prices. World prices are reported for international transactions only and equal market prices plus international transport margins. International transport margins can be delivered to a country other than either the importing or exporting region (e.g., a Swedish company transporting goods from Denmark to Norway). Net taxes or subsidies are reported as the difference between agent and market prices (or world prices, when international transactions are involved).

The matrix of economic transactions, **T**, can be decomposed as $\mathbf{T} = \mathbf{T}^L + \mathbf{T}^I$, where $\mathbf{T}^L$ and $\mathbf{T}^I$ are matrices of *domestic* and *international* transactions, respectively. In $\mathbf{T}^L$ all entries corresponding to international transactions are equal to zero and in T^I entries corresponding to domestic transactions are in principle equal to zero, but they can be non-zero if the region is an aggregation of a group of nations.

The construction of T^I involves estimation and is discussed in subsection 5.2.3. In the next subsection we address the construction of T^L, e^L, x, y and v, i.e. all domestic data.

5.2.2 *Domestic data*

The MRIO model presents flows in market prices, since those reflect the true monetary flow from buyer to seller, and the difference between market and agent price is a payment to or from the government. Therefore, the entries of T^L are obtained from market price GTAP tables (using the GTAP codes): VDFM (inter-industry domestic flows); VDPM (flows from industries to private households); VDGM (flows from firms to government); column CGDS of table VDFM (flows from firms to investment); VFM (flows from endowments to firms).

The differences between agents' and market prices are taxes (if positive) and subsidies (if negative) and are obtained by subtracting each of the above tables from the corresponding table in agents' prices: (VFM - EVOA) are endowment taxes; (VDPA - VDPM), (VDGA - VDGM) and the columns corresponding to CGDS of (VDFA - VDFM) are the final demand taxes (respectively, paid by the household, government and investment sectors).

For reasons discussed in Subsection 5.2.3, imports for final demand were merged with domestic final demand, and so the following adjustments were necessary. To the vectors VDPM, VDGM and VDFM(CGDS) (transactions from domestic industries to households, government and investment) were added, respectively, VIPM, VIGM and VIFM(CGDS). To the taxes paid by final demand were added, respectively, (VIPA - VIPM), (VIGA - VIGM) and (VIFA(CGDS) - VIFM(CGDS)).

Industries are more complex with regard to this subject, because they pay different types of taxes/subsidies (production, export, import, profit, etc.), while the GTAP only reports net taxes/subsidies. Furthermore, the aggregation level at which tax data is reported is not always consistent. Thus, we decided to assign to industries the taxes paid over endowment factors (the difference between EVFA and VFM), which already include a substantial fraction of total taxes, and the remainder of firms' taxes appears as a balancing item. We shall now explain the balancing procedure.

Total output of firms was obtained as the sum of total domestic output (i.e. row sums of T^L) and the vector of international output t^r. Total input of firms was obtained as the sum of total domestic input (row columns of T^L) and international input, t^c. (t^c and t^r are described in the following section.)

The largest value of total input or output was assigned to vector **x**, and the difference was added to taxes or subsidies (to ensure all transactions are positive). This procedure balanced the industry entries of **x**.

The final demand, endowment, tax and subsidy entries of **x** were taken as the maximum of the corresponding row or column sums of $\mathbf{T}^L$, and vectors **y** and **v** were obtained as the difference between the row or column sums of $\mathbf{T}^L$ and the corresponding **x** entry. This balancing procedure always yielded the qualitatively correct result (final demand and subsidies delivering goods and services to regional households in **y**, and endowments and taxes delivering payments to regional households in **v**).

The GTAP database is criticized because its source data is not from a single base year, some tables being somewhat old, and this source data is then substantially processed, with no reporting of the resulting errors (Peters and Hertwich 2008a, Wiedmann *et al.* 2008b). Still, we decided to use the GTAP data because it has a more extensive coverage than the alternatives, such as the OECD database (Yamano and Ahmad 2006) and requires less effort than collecting individual tables from national statistics offices, which involves additional processing, such as sector reclassification and exchange-rate conversions.

5.2.3 *International data*

The GTAP database distinguishes between intermediate imports (inputs of industries) and final imports (imports for final demand) and also reports partial data on margins for international transportation. However, we believe that the additional complexity required to maintain this distinction would not pay off for the refinement in results it would provide. Thus, we consider that imports to the final demand of a given sector are first imported by the domestic production sector and then delivered to final demand by the domestic sector. We also consider that import taxes and trade margins (which are paid by the buyer and received by third parties) are received by the seller (thus import taxes end up being paid by the exporting region). These assumptions induce systematic errors, but it would be difficult to proceed otherwise, given the structure of the data.

Under these assumptions, the total input from international trade of a production sector, $\mathbf{t}^c$, is obtained as the sum of intermediate import (VIFM) and import for final demand (VIFM(CGDS), VIPM and VIGM). The total output for international trade of a firm sector, $\mathbf{t}^r$, is obtained from table VIMS. The vectors of the column and row sums of the matrix of international flows $\mathbf{T}^I$ are $\mathbf{t}^c$ and $\mathbf{t}^r$.

Disaggregate data on international trade (the entries of $\mathbf{T}^I$) is not available. However, the GTAP database provides two sets of complementary constraints on these flows: imports by industries, discriminated by importing region and sector and by exporting sector (but not exporting region); and bilateral trade at sector level, discriminated by exporting region and sector and by importing region (but not by importing sector).

For the sake of clarity, in the remainder of this section we shall use t_{ij}^{ab} to describe the flow from sector i of region a to sector j of region b, $*$ to denote sum over all values, ex and im to denote respectively export and import. Therefore, imports by firms is a set of values im_{ij}^{*b}, obtained from table VIFM (where to entries im_{ii}^{*b} were added the corresponding values from VIFM(CGDS), VIPM and VIGM, as explained in the first paragraph of this subsection), and bilateral trade is a set of values ex_{i*}^{ab}, obtained from table VIMS.

In the next section we show how $\mathbf{t^r}$, $\mathbf{t^c}$ and the im_{ii}^{*b}'s and ex_{i*}^{ab}'s, can be used to construct $\mathbf{T^I}$.

5.3 Computation

5.3.1 Estimation of international flows

The standard approach to construct table $\mathbf{T^I}$ is the 'trade share' method, which consists of using a region's share of total global exports to disaggregate imports on a sector basis (Lenzen *et al.* 2004, Peters and Hertwich 2006). Using the present notation it is $t_{ij}^{ab} = im_{ij}^{*b} \frac{ex_{**}^{a*}}{ex_{**}^{**}}$.

This method ensures that $t_{ij}^{*b} = im_{ij}^{*b}$ (i.e. the sum of disaggregated imports over exporting region matches the known import constraint) and is usually used when detailed bilateral trade data is not available. The 'trade share' method is considered a better option than the alternatives employed in the lack of more detailed data (either to ignore international indirect effects or to consider that foreign sectors have intensities identical to domestic sectors (Munksgaard and Pedersen 2001)), but worse than having detailed 'survey' transactions data (Ackerman *et al.* 2007).

To make full use of the detailed data provided by the GTAP, two 'trade share'-type methods are possible: to use bilateral data to disaggregate imports or to use import data to disaggregate bilateral trade. These are respectively:

$$t_{ij}^{ab} = im_{ij}^{*b} \frac{ex_{i*}^{ab}}{ex_{i*}^{*b}} \tag{5.1}$$

$$t_{ij}^{ab} = ex_{i*}^{ab} \frac{im_{ij}^{ab}}{im_{i*}^{*b}} \tag{5.2}$$

We also consider two more aggregated estimation methods, in its export and import share versions:

$$t_{ij}^{ab} = im_{ij}^{*b} \frac{ex_{**}^{a*}}{ex_{**}^{**}} \tag{5.3}$$

$$t_{ij}^{ab} = ex_{i*}^{ab} \frac{im_{*j}^{**}}{im_{**}^{**}} \tag{5.4}$$

We use these four different estimation methods to construct the matrix of international trade $\mathbf{T^I}$. We insert $\mathbf{T^I}$ in the full matrix of inter-sectoral transactions $\mathbf{T}$ and balance the latter using the RAS method. (The RAS method is a technique that alters the entries of a square matrix to ensure that row and column sums match. More information on this technique can be found in UN [1999].)

Each of these methods provides a different estimate for t_{ij}^{ab}. We computed the carbon intensities, $\mathbf{m^U}$ and $\mathbf{m^D}$, and embodied emissions, $\mathbf{e^U}$ and $\mathbf{e^D}$, using the different estimates. We considered the arithmetic mean of the computed quantities as the best estimate and the largest difference between the extreme values and the mean as the error introduced by the estimation of international inter-industry transactions.

This is different from a Monte Carlo error estimation procedure, where international inter-industry transactions are sampled from independent unimodal distributions. The Monte Carlo approach leads to cancelling out effects, and thus the region of the state space around the expected value is more strongly sampled than more distant regions (Wiedmann *et al.* 2008a). Using our approach, the several international inter-industry transactions are not estimated independently, and the weight assigned to each estimation procedure 5.1 to 5.4 is the same. Therefore, our approach functions as a sensitivity analysis, exploring in a more balanced way the full range of potential values of international inter-industry transactions.

IO models, of course, are subject to many other error types. But these errors equally affect a large MRIO as presented here, and more conventional IO models, and as they have already been addressed in the literature (Lenzen 2001, Wiedmann *et al.* 2008a), we do not examine them here.

5.3.2 Domestic indirect emissions

Total embodied emissions can be decomposed into three fractions:

$$m_i^U = m_i^{Ui} + m_i^{Ud} + m_i^L \tag{5.5}$$

$$m_i^D = m_i^{Di} + m_i^{Dd} + m_i^L \tag{5.6}$$

In the previous expressions, m_i^{Ui} and m_i^{Ud} are, respectively, *upstream international indirect carbon intensity* and *upstream domestic indirect carbon intensity*; m_i^{Di} and m_i^{Dd} are, respectively, *downstream international indirect carbon intensity* and *downstream domestic indirect carbon intensity*; and m_i^L is *direct carbon intensity*.

Direct carbon intensity, defined in Eq. 2.11, indicates how much of the total emissions embodied in a flow results from the direct emissions occurring in the producing/consuming sector (respectively for upstream/downstream emissions). Indirect carbon intensity (total minus direct intensity) can be decomposed into a domestic and an international fraction, which are distinguished by the location of the source of direct emissions (whether within national borders or outside them). Total domestic intensities (direct plus indirect) are computed by assigning zero intensity to the corresponding international flows (imports when computing upstream embodied emissions and exports when computing downstream embodied emissions), and we computed all such quantities simultaneously by removing international trade from the IO model (i.e. we set $\mathbf{T^I}$ equal to a null matrix and added $\mathbf{t^r}$ and $\mathbf{t^c}$ to the outputs to subsidies and inputs from taxes, respectively, as those sectors do not have direct emissions).

The discrimination of direct and indirect domestic and international emissions is important for two reasons. First, it informs whether it is necessary to account for indirect effects at all (it is, if the ratio of direct to total intensities is small). Second, it informs whether it is necessary to account for international indirect effects (if the ratio of international to total intensities is large). These are important questions, because it is easier to account only for direct effects – discarding all of the IO apparatus; and even if indirect effects are important, it is easier to account only for domestic indirect effects – discarding the multi-regional component of the IO model.

We shall compare the fraction of international to total intensities with the error margin, because if the former is smaller than the latter, it implies that the MRIO model is not informative, and a simpler national IO model should be used instead.

5.3.3 Computational procedure

The following steps were taken to compute embodied emissions and carbon responsibilities.

The GTAP data was extracted from the database using two freely available applications: the gdxtohar software, available at http://www.mpsge.org/gdxhar/index.html, to convert from har to gdx format; the gdxviewer software, available at www.gams.com/download, to convert from gdx to ASCII format.

The data was further processed using the C or Octave languages to construct the full MRIO model.

The computation of carbon intensities 2.6 and 2.10, was performed for the several scenarios reported (the estimation methods mentioned in 5.3.1 and the domestic total emissions mentioned in 5.3.2), using a linear system solver from the LAPACK linear algebra library.

The treatment of the results was also performed using C, Octave and OpenOffice.

In the following chapters we report the results.

6 Carbon responsibility of world regions

6.1 Role of international trade

In this chapter we report the results of consumer and producer responsibility of eighty-seven world regions computed using a MRIO model and using the GTAP database as source data. The basic theory underlying these calculations is presented in Chapter 2, the definition of responsibilities is presented in Chapter 4 and the description of the source data and computations is presented in Chapter 5.

However, before presenting more detailed results, we use the MRIO model to provide empirical answers to the following questions: (1) whether it is worth accounting for indirect effects; (2) whether it is worth accounting for international indirect effects; (3) whether current techniques suffice to provide accurate results.

The first question addresses the relevance of IO analysis itself: if indirect effects are only a small fraction of total carbon intensities, it means that by using only direct emissions it is already possible to have a good idea of the carbon emissions associated with economic activities. If this is the case, IO analysis is not very useful.

The second question addresses the relevance of building a multi-regional IO model: if international indirect effects are only a small fraction of total carbon intensities, it means that using only a national IO table is preferable, as it is both easier to obtain results and involves less uncertainty.

The third question addresses the relevance of estimating international inter-industry transactions. Such quantities need to be estimated, on the basis of more aggregate quantities, and if the error introduced by the estimation (as a fraction of total carbon intensity) is larger than the international indirect effects (as a fraction of total carbon intensity) it means that even though an MRIO model is indeed necessary to provide a good picture of total embodied emissions, it is not acceptable to estimate international inter-industry transactions.

We answer these questions statistically. For a total of 87 regions × 57 sectors = 4959 data points of both upstream and downstream carbon intensities we recorded three ratios: the fraction of indirect effects, the fraction of international indirect effects and the fraction of estimation errors (see Sections 5.3.2 and 5.3.1 for the meaning of these quantities).

For every data point i, and for both upstream and downstream emissions, the fraction of indirect effects is the ratio of the sum of domestic carbon intensity and international carbon intensity to total carbon intensity:

$$\frac{m_i^{Ud} + m_i^{Ui}}{m_i^U} \quad \text{and} \quad \frac{m_i^{Dd} + m_i^{Di}}{m_i^D}$$

The fraction of international indirect effects is the ratio of international carbon intensity to total carbon intensity:

$$\frac{m_i^{Ui}}{m_i^U} \quad \text{and} \quad \frac{m_i^{Di}}{m_i^D}$$

The fraction of estimation errors is the ratio of the estimation error to total carbon intensity:

$$\frac{\epsilon_i^U}{m_i^U} \quad \text{and} \quad \frac{\epsilon_i^D}{m_i^D}$$

(The explanation of how the estimation error is computed is reported in Section 5.3.1.)

For each of these six ratios (which are bound by 0 and 1), we treat the 4959 data points as a sample from a probability distribution, whose accumulated probability functions are depicted in the following figures. Figure 6.1 reports the fraction of indirect effects (domestic and international) on total intensities. Figure 6.2 reports the fraction of international indirect effects on total intensities. Figure 6.3 reports the ratio of estimation error to total intensity.

Let X denote a random variable. These figures represent y, the probability that the random variable (a randomly chosen data point) is smaller than x, $P(X < x) = y$ (where x and y are depicted in the x-axis and y-axis respectively). Since $0 \leq X \leq 1$, $P(X < 0) = 0$ and $P(X < 1) = 1$, the accumulated probability function is a line that crosses the points (0,0) and (1,1). Remember also that if $P(X < x) = y$, then $P(X \geq x) = 1 - y$.

Let X represent the fraction of upstream indirect effects. Figure 6.1 shows that $P(X < 50\%) = 15\%$. That is, for 15% of the data points, indirect effects represent less than 50% of total upstream carbon intensity. That implies

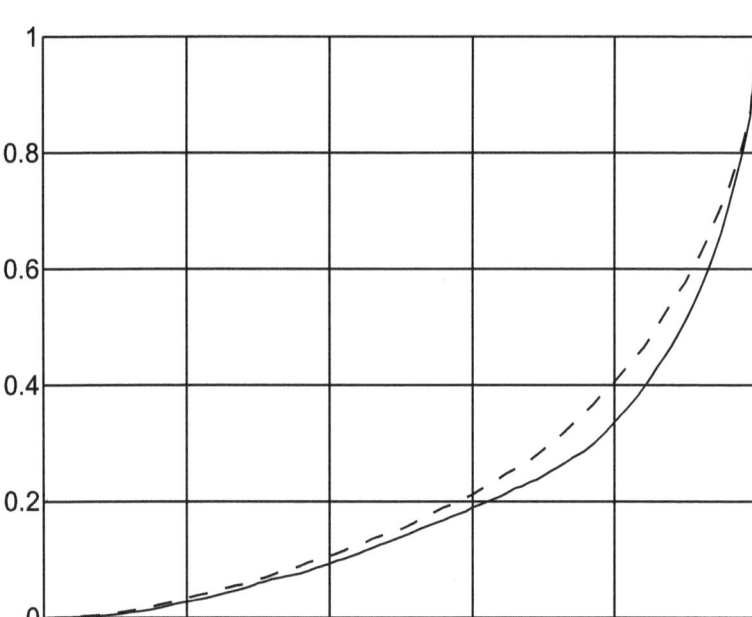

Figure 6.1 Accumulated probability functions for indirect effects in upstream (full line) and downstream (dashed line) sectoral intensities.

that for 85% of the data points, indirect effects represent more than 50% of total upstream carbon intensity. Thus, we consider that, without any doubt, it makes sense to take indirect effects into account.

Figure 6.1 also shows that downstream indirect effects are less important (as a fraction of total downstream intensity) than upstream indirect effects (as a function of total upstream intensity), but the difference is small.

Now let X represent the fraction of upstream international indirect effects. Figure 6.2 shows that $P(X < 50\%) = 60\%$. That is, for 40% of the data points, international indirect effects represent more than 50% of total upstream carbon intensity. Thus, international indirect effects are less important than domestic indirect effects, but we still consider that it makes sense to take international indirect effects into account.

Figure 6.2 also shows that downstream international indirect effects are less important (as a fraction of total downstream intensity) than upstream international indirect effects (as a function of total upstream intensity), and this difference is meaningful. If we let X represent the fraction of downstream international indirect effects, Figure 6.2 shows that $P(X < 50\%) = 80\%$,

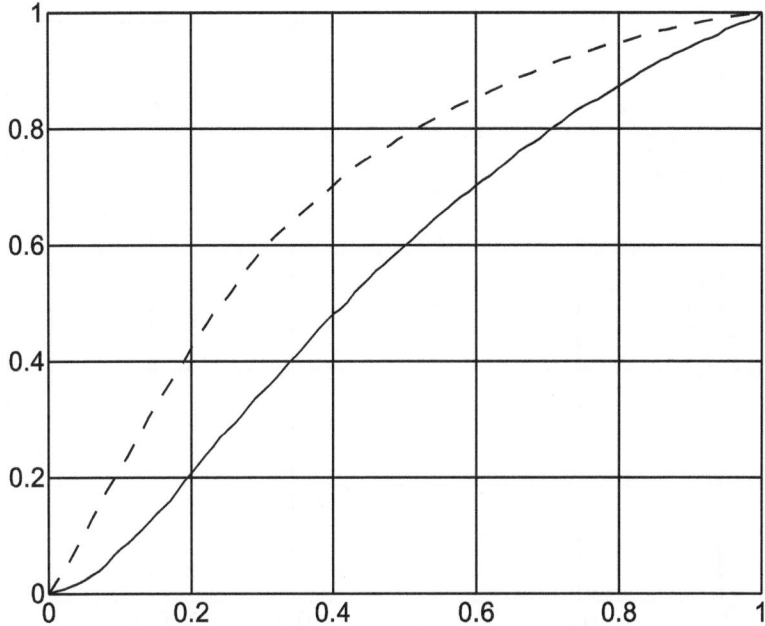

Figure 6.2 Accumulated probability functions for international indirect effects in upstream (full-line) and down stream (dashed line) sectoral intensities.

which implies that for only 20 % of the data points international indirect effects represent more than 50%. Thus, international indirect effects are less relevant for downstream than for upstream embodied emissions.

Finally, let X represent the error fraction of upstream international indirect effects. Figure 6.3 shows that $P(X < 20\%) = 90\%$ and $P(X < 10\%) = 80\%$. That is, for 90% of the data points, the estimation error is smaller than 20% and for 80% of the data points, the estimation error is smaller than 10%. Thus, the error fraction is well below the fraction of international indirect effects, for the large majority of data points. We thus consider that the estimation error is acceptable and that even with the uncertainties of estimating international inter-industry transactions, a multi-regional IO model is required to have a good picture of total upstream carbon intensities.

Figure 6.3 also shows that even though the accumulated probability function of the upstream error ratio is larger than the accumulated probability function of the downstream error ratio, the difference between the two curves is not meaningful.

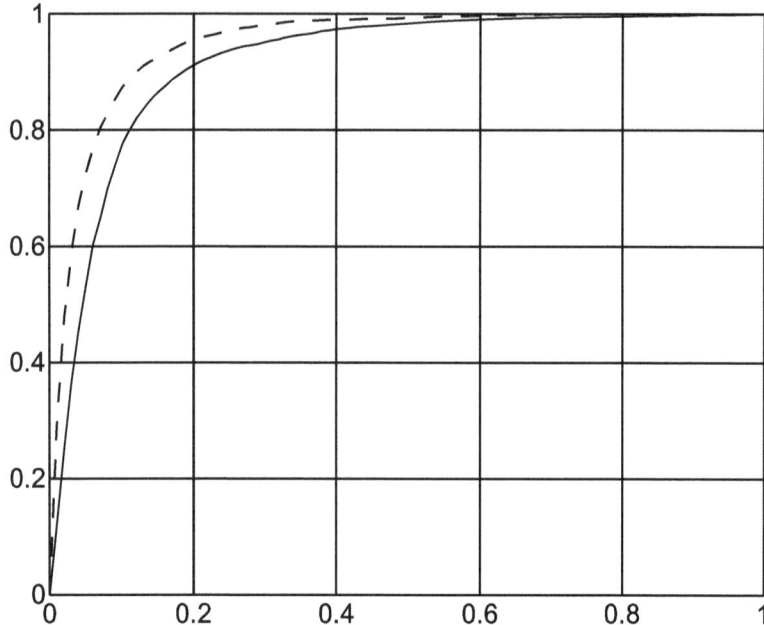

Figure 6.3 Accumulated probability functions for the ratio of estimation error to total
upstream (full line) and downstream (dashed line) sectoral intensities.

Thus, in brief, we consider that in order to account for both the total
upstream and downstream embodied emissions of the economic sectors
of the full GTAP database, it is necessary to use a full MRIO model.
However, many of the countries present in the GTAP database are small,
open economies. For a large country, for which international trade is less
important, a domestic IO table might suffice.

6.2 Responsibilities vs GDP

In this section we examine how direct emissions, consumer responsibility
and producer responsibility are related to gross domestic product (GDP).
GDP, a commonly used metric of economic welfare, is defined as domestic
final consumption plus exports minus imports, and is equivalent to gross
domestic income (GDI), the primary inputs of the IO model.

Figures 6.4, 6.5 and 6.6 display, respectively, per capita direct emissions,
consumer responsibility and producer responsibility against per capita GDP.
All figures are in a log-log scale, and display the 87 GTAP region data points.

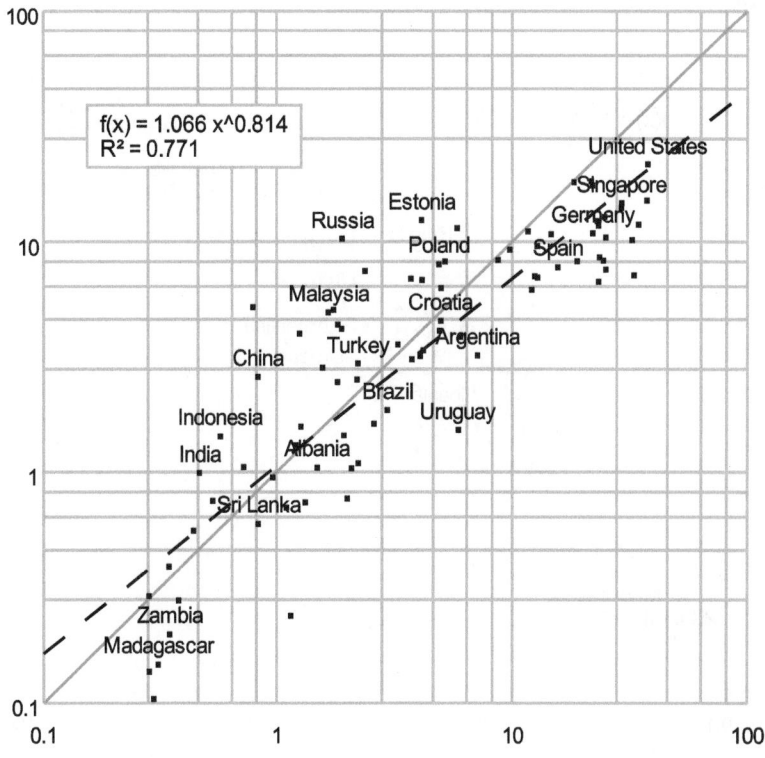

Figure 6.4 Per capital direct emissions (tonCO$_2$) vs per capita GDP (10^3 USD) for all GTAP regions.

Figures 6.4, 6.5 and 6.6 also display linear regressions in a log-log scale, implying that the data can be described by a power-law of the type $y = ax^b$. An increase in GDP is associated with an increase in all types of emissions (positive coefficients a and b). All regressions have good fit values (high R^2) but per capita consumer responsibility displays the best fit, with $R^2 = 0.8606$. The coefficient b is the elasticity of emissions with respect to GDP. In all cases emissions increase strongly with GDP, but it is producer responsibility that has a higher elasticity with $b = 0.83$. This value ($b = 0.83$) is in accordance with values reported by Hertwich and Peters (2009). An elasticity of 0.83 means that, for a 1% increase in GDP, producer responsibility increases by 0.83. It should also be noted that in all cases the elasticity is smaller than 1, implying that an additional unit of GDP always leads to less than one additional unit of carbon emissions.

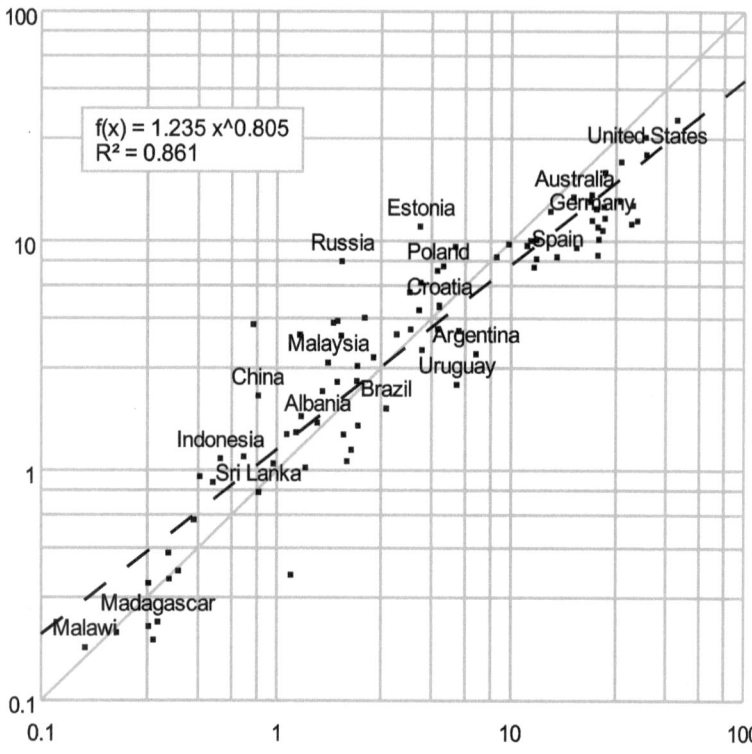

Figure 6.5 Per capita consumer responsibility (tonCO_2) vs per capita GDP (10^3 USD) for all regions.

Some of the datapoints in Figures 6.4, 6.5 and 6.6 are labelled, showing that GTAP regions cluster around world regions. The cluster on the right is the group of economies with higher per capita GDP values (mainly OECD countries), in the centre there is a cluster of middle income countries (Eastern European Countries, Latin America and Asia) and on the left are the countries with lowest per capita GDP (Africa). In the remainder of this chapter we shall study these six aggregated world regions. Table A.5 in the appendix lists the GTAP regions that compose each aggregated world region.

Now we can also examine how direct emissions, consumer and producer responsibilities are related to GDP at the level of world regions. We performed linear regressions to the log-transformed regional data sets, and obtained the elasticity and R^2 values reported in Table 6.1.

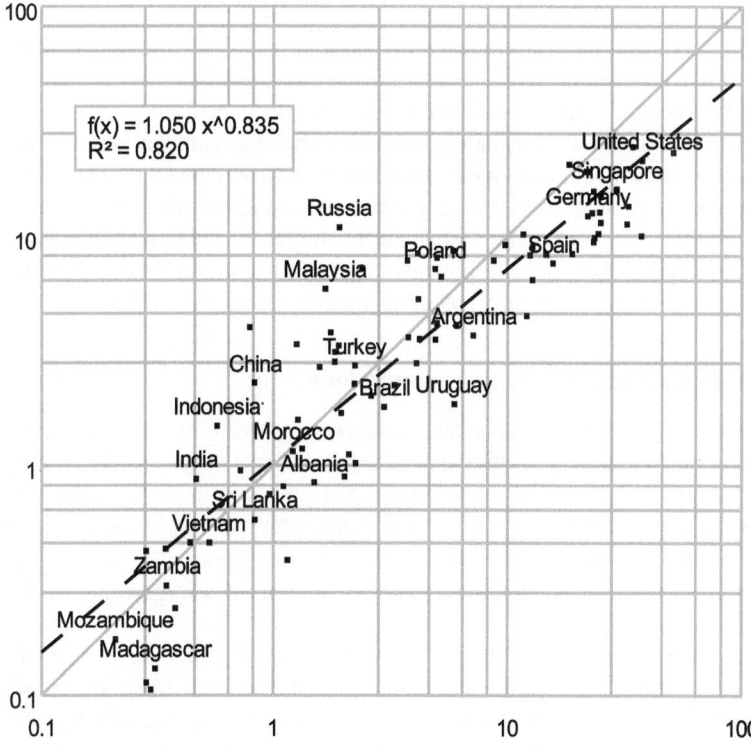

Figure 6.6 Per capita producer responsibility (tonCO$_2$) vs per capita GDP (10^3 USD) for all regions.

Table 6.1 World region elasticities of direct emissions, consumer and producer responsibility with respect to GDP. R^2 is reported inside brackets

Elasticity	Direct	Consumer	Producer
Africa	1.26	1.14	1.17
	(0.90)	(0.96)	(0.87)
Asia	0.92	0.84	0.95
	(0.65)	(0.91)	(0.84)
Developed	0.64	0.80	0.78
	(0.15)	(0.64)	(0.61)
Latin America	0.74	0.67	0.68
	(0.54)	(0.64)	(0.63)
Eastern Europe	0.52	0.78	0.95
	(0.28)	(0.64)	(0.60)
Fossil Fuel E.	0.56	0.62	0.65
	(0.63)	(0.76)	(0.75)

The higher elasticities are reported for Africa, whereas Fossil Fuel Exporters have the lowest. Generally speaking, regions with higher average GDP per capita have lower elasticity, implying that with economic development, all types of carbon emissions are decoupled from GDP.

The elasticities of Africa are greater than one, implying that (in a cross-country comparison) carbon emissions increase faster than GDP. This finding is not surprising, since these economies are at an early stage of economic development, dominated by obsolete technologies. In addition, at this stage of development governments policies are generally more aimed at economic development than environmental protection (Azomahou *et al.* 2006).

6.3 Responsibilities vs direct emissions

We computed the consumer and producer responsibility of all regions, and compared these quantities against regional direct emissions. Table A.6 in Section A.3 reports absolute direct emissions, consumer and producer carbon responsibilities (E_k^L, E_k^U and E_k^D) and error margins for each region k, for the year 2001 (MtonCO$_2$ or 10^9 kgCO$_2$). Figure 6.7 shows these quantities for selected countries, representative of the six aggregate world regions.

In Table A.6 we see that, for the purpose of computing regional responsibilities, the errors associated with international trade are small ($< 5\%$ for most regions). However, these errors only account for the uncertainty affecting the estimation of international trade, as described in Section 5.3.1.

Absolute values can be normalized to per capita values, giving a clearer indication of the environmental performance of a country and also enabling inter-country comparisons. Table A.7 in Appendix Section A.3 reports the per capita (p.c.) direct emissions, consumer and producer carbon responsibilities (p.c. E_k^L, p.c. E_k^U and p.c. E_k^D) and error margins for every region k (tonCO$_2$ p.c.). Figure 6.8 shows per capita direct emissions and responsibilities for countries that are representative of each of the six world regions mentioned in Section 6.2.

Responsibilities deviate little from direct emissions for large countries, but for small and poor countries the deviations can be substantial. Large countries tend to be more closed in economic terms, with much national trade, while small countries are more dependent on international trade. In Figure 6.8 we see that the countries with higher per capita responsibilities are the United States and Switzerland (Developed Economies). Despite a similar GDP, the relationship between responsibilities and direct emissions is different for both countries. Switzerland's consumer and producer

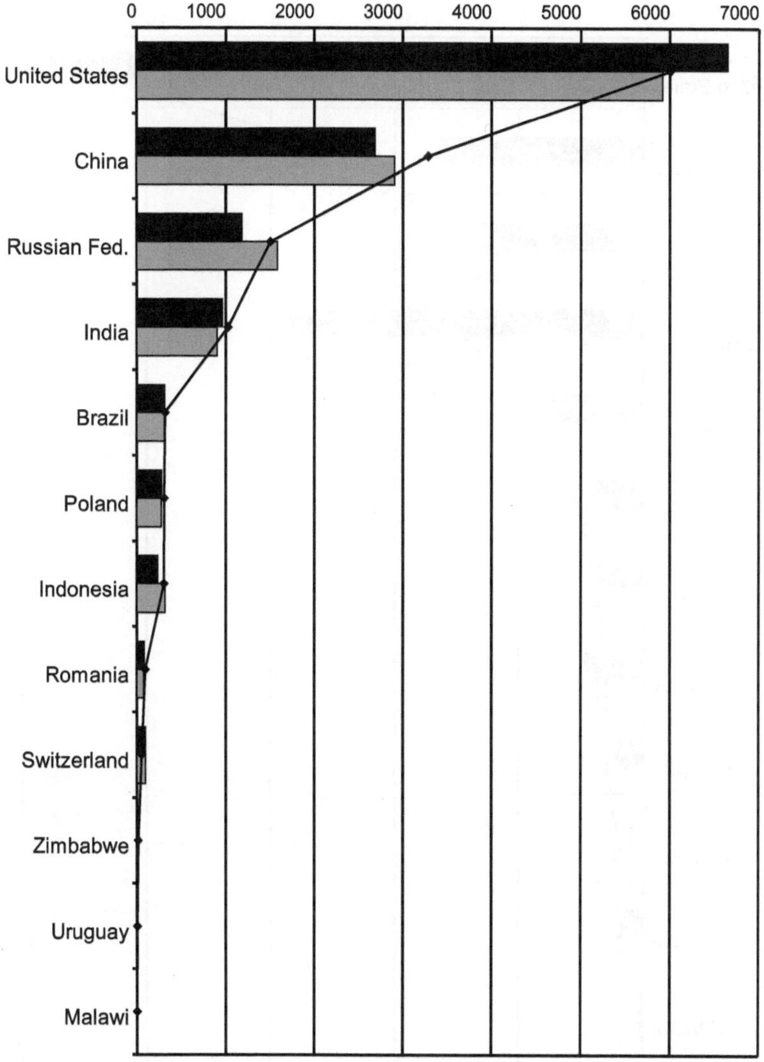

Figure 6.7 Absolute direct emissions (dotted line), consumer (black bars) and producer (grey bars) responsibility of selected countries (MtonCO$_2$).

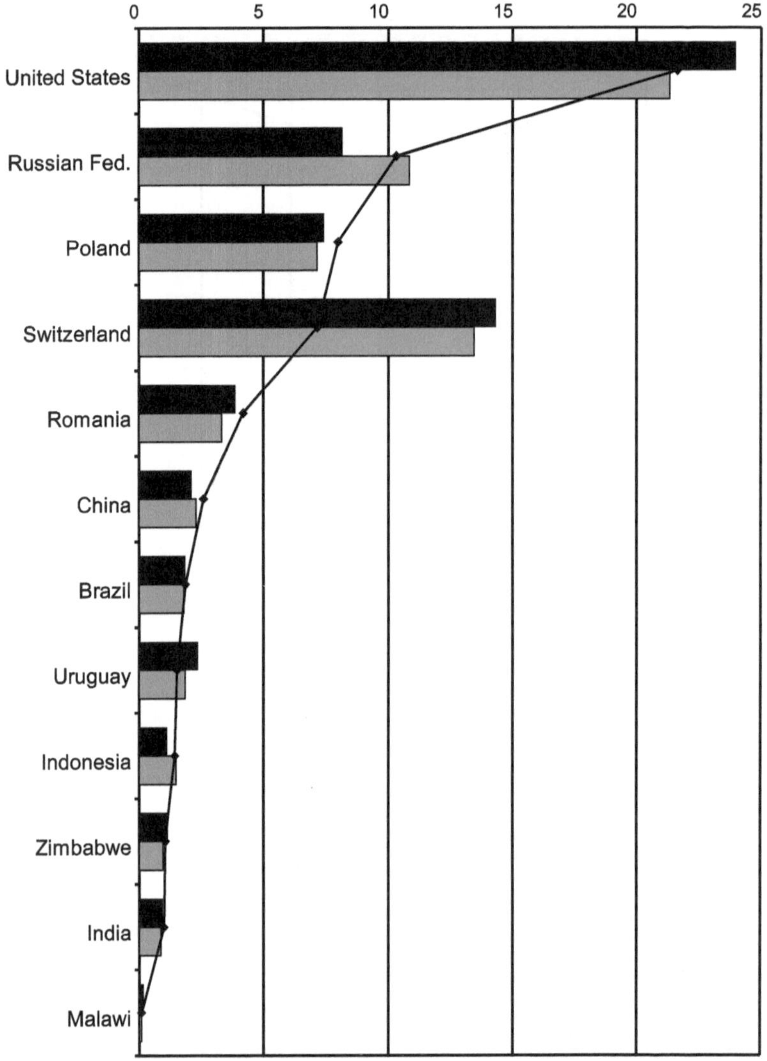

Figure 6.8 Per capita direct emissions (dotted line), consumer (black bars) and producer (grey bars) responsibility of selected countries (tonCO$_2$ p.c.).

responsibility are, respectively, 100% and 88% higher than direct emissions. Regarding the United States, consumer responsibility is 11% higher than direct emissions whereas producer responsibility is 5% lower. Regarding the weight of international trade, Switzerland's imports are 43% of GDP and exports are 48%, while United States' imports represent 13% of GDP and exports 9%.

The Russian Federation (a Fossil Fuel Exporter) has producer responsibility and direct emissions of the same magnitude and higher than consumer responsibility. The high direct emissions of the Russian Federation reflect low domestic gas and electricity prices (IEA 2003). Eastern European countries have high values of direct emissions due to a high dependence on coal for electric power generation and the lack of environmental regulations associated with coal usage (IEA 2003). Comparatively, the other countries presented have low values of per capita direct emissions and responsibilities (below 5 tonCO$_2$ p.c.).

The absolute and per capita responsibilities of the aggregated regions are reported in Tables 6.2 and 6.3 and in Figures 6.9 and 6.10.

Table 6.2 Absolute direct emissions, consumer and producer responsibility of aggregated regions (MtonCO$_2$), ordered by decreasing absolute direct emissions

Region	E_k^L	E_k^U	E_k^D
Developed	11228.48	12797.80	11645.61
Asia	5549.12	4883.19	4913.24
Fossil Fuel Exporters	4871.60	4039.51	5368.95
Eastern Europe	1473.74	1308.53	1187.05
Latin America	1146.26	1203.17	1148.62
Africa	481.00	518.00	486.74

Table 6.3 Per capita direct emissions, consumer and producer responsibility of all aggregated regions (tonCO$_2$ p.c.), ordered by decreasing per capita direct emissions

Region	p.c. E_k^L	p.c. E_k^U	p.c. E_k^D
Developed	13.55	15.45	14.06
Fossil Fuel E.	5.91	4.90	6.52
Eastern Europe	5.90	5.24	4.75
Latin America	2.50	2.62	2.50
Asia	1.83	1.61	1.62
Africa	0.65	0.70	0.66

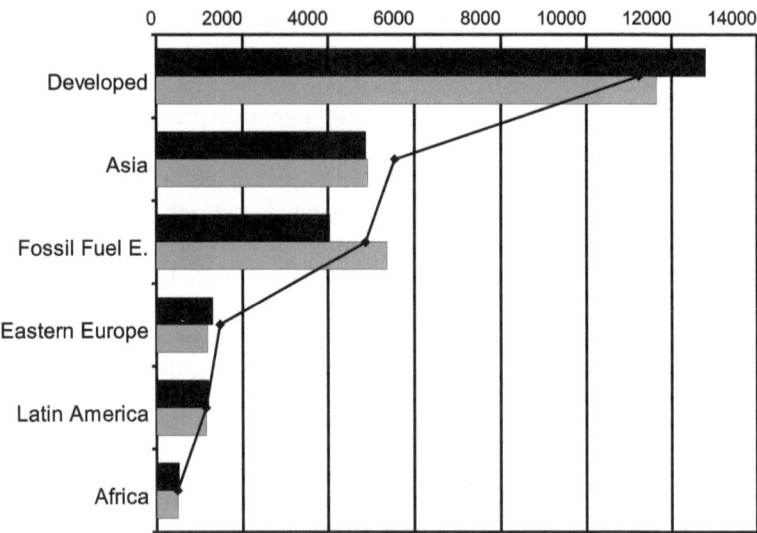

Figure 6.9 Absolute direct emissions (dotted line), consumer (black bars) and producer (grey bars) responsibility of aggregated regions (MtonCO$_2$).

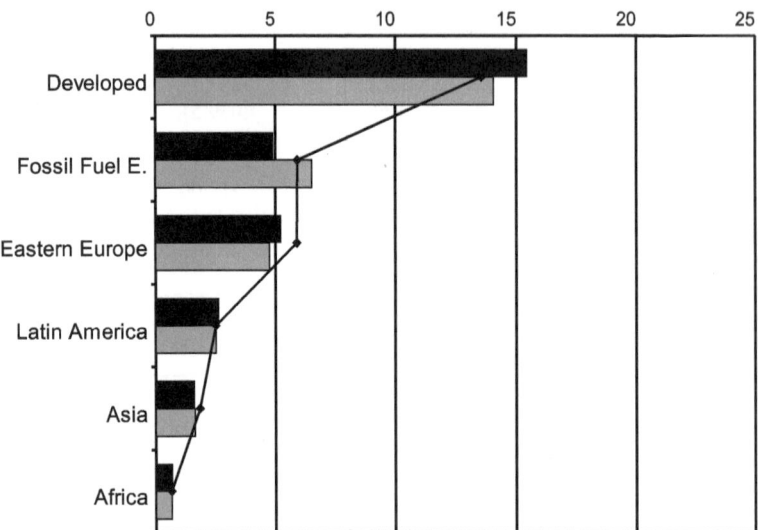

Figure 6.10 Per capita direct emissions (dotted line), consumer (black bars) and producer (grey bars) responsibility of aggregated regions (tonCO$_2$ p.c.).

Concerning absolute values (Figure 6.9), we see that almost 90% of world total emissions are from Asia, Developed Economies and Fossil Fuel Exporters. Regarding per capita emissions (Figure 6.10), Developed Economies exhibit values three times larger than the next highest emitting regions (Fossil Fuel Exporters and Eastern Europe), which in turn exhibit values at least the double of the remaining regions.

The relation between responsibilities and direct emissions is different for the different regions: Developed Economies have both consumer and producer responsibility higher than direct emissions; Asian and Eastern European countries have both responsibilities lower than direct emissions; Fossil Fuel Exporters have consumer responsibility lower and producer responsibility higher than direct emissions; Latin America and Africa do not display noticeable differences between either of those quantities. These patterns are apparent in Figure 6.12 and their explanation lies in the carbon content of bilateral trade, which we study in the following Section.

6.4 Carbon trade balance

As we saw in Section 4.2.4, the consumer carbon responsibility of region k, E_k^C, is the sum of the total upstream emissions embodied in the final demand of that region, $\sum_{i \in S_k} e_{i0}^U$. Likewise, the producer carbon responsibility of region k, E_k^P is the sum of the total downstream emissions embodied in the primary inputs of that region, $\sum_{i \in S_k} e_{0i}^D$.

However, we also know that total embodied emissions obey conservation laws (Eqs. 2.4 and 2.7), stating that (for upstream embodied emissions) what goes out, whether to final demand or exports, matches direct emissions plus what comes in in imports (and the converse for downstream embodied emissions). Thus, we can write the following balances to the total upstream and downstream embodied emissions of region k:

$$\sum_{i \in S_k} e_{i0}^U + \sum_{k' \in R_W, k' \neq k} \sum_{i \in S_k} \sum_{j \in S_k'} e_{ij}^U = \sum_{i \in S_k} e_i^L + \sum_{k' \in R_W, k' \neq k} \sum_{i \in S_k} \sum_{j \in S_k'} e_{ji}^U$$

$$\sum_{i \in S_k} e_{0i}^D + \sum_{k' \in R_W, k' \neq k} \sum_{i \in S_k} \sum_{j \in S_k'} e_{ji}^D = \sum_{i \in S_k} e_i^L + \sum_{k' \in R_W, k' \neq k} \sum_{i \in S_k} \sum_{j \in S_k'} e_{ij}^D$$

The previous expressions can be simplified to:

$$E_k^C + E_{kW}^U = E_k^L + E_{Wk}^U$$
$$E_k^P + E_{Wk}^D = E_k^L + E_{kW}^D$$

In the previous set of equations we used subscripts kW and Wk to denote, respectively, total export from region k to the rest of the world and imports to region k from the rest of the world. Thus, in words, the previous set of equations reads, for any given region:

Consumer responsibility + Total upstream emissions embodied in exports = Direct emissions + Total upstream emissions embodied in imports,

Producer responsibility + Total downstream emissions embodied in imports = Direct emissions + Total downstream emissions embodied in exports.

Rearranging these expressions we obtain:

$$b_k^U \equiv E_k^C - E_k^L = E_{Wk}^U - E_{kW}^U \tag{6.1}$$

$$b_k^D \equiv E_k^P - E_k^L = E_{kW}^D - E_{Wk}^D \tag{6.2}$$

In Eqs. 6.1 and 6.2 we have introduced two new quantities: *upstream carbon trade balance* (UCTB), b_k^U, and *downstream carbon trade balance* (DCTB), b_k^D.

UCTB is the difference between consumer responsibility and direct emissions, and is also the difference between the total upstream carbon emissions embodied in imports and exports. Thus, if $b_k^U > 0$, it means that region k is receiving more upstream embodied carbon emissions, TUECEF, in its imports than disposing of them in its exports. That is, such a region is a net *importer* of total upstream embodied carbon emissions.

DCTB is the difference between producer responsibility and direct emissions, and is also the difference between the total downstream embodied carbon emissions, TDECEF, embodied in exports and imports. Thus, if $b_k^D > 0$, it means that region k is receiving more downstream embodied carbon emissions, TDECEF, in its exports than disposing of them in its imports. That is, such a region is a net *importer* of total downstream embodied carbon emissions.

In the following analysis we study per capita UCTB and DCTB, since per capita values make inter-regional comparisons easier.

Table A.8 in Appendix Section A.3 reports the carbon trade balances for all GTAP regions and Figure 6.11 displays the carbon trade balances for the subset of GTAP regions whose carbon trade balances are closer to the origin. Figure 6.11 shows that the GTAP regions tend to cluster into world regions according to criteria of geography and, more importantly, of economic structure.

Table 6.4 and Figure 6.12 show the carbon trade balances of the six aggregated world regions. In the Northeast quadrant of Figure 6.12 we find developed economies (D); in the Southwest quadrant we find Eastern

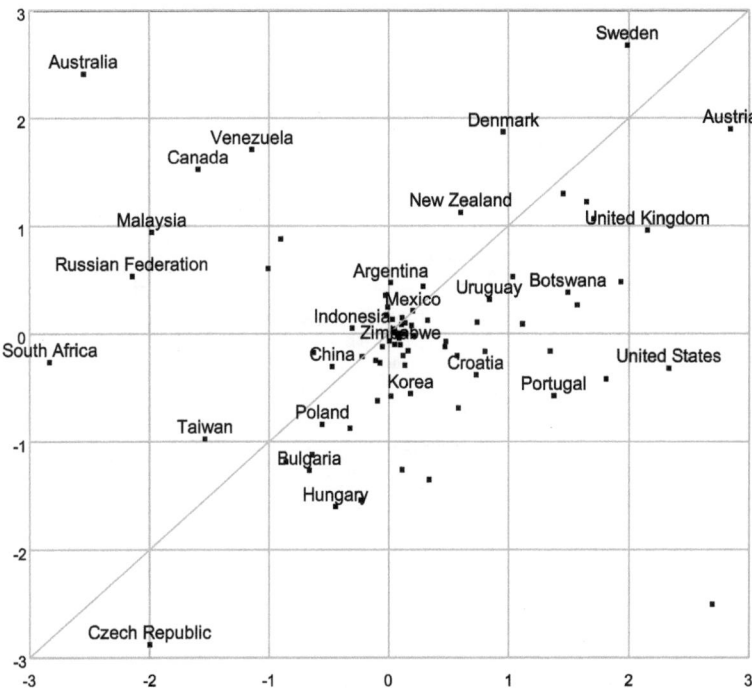

Figure 6.11 Upstream (x-axis) vs downstream carbon trade balance (y-axis) for GTAP regions (tonCO$_2$ per capita).

European (EE) countries; and in the Northwest quadrant we find the main fossil fuel exporters (F) (IEA 2004). Close to the origin we find Asian (A), African (AF) and Latin American (L) countries, with Asia in the Southwest quadrant and both Africa and Latin America placed East of the y-axis.

Table 6.4 Upstream and downstream carbon trade balances for six aggregated world regions (tonCO$_2$ per capita)

Region	b_k^U	b_k^D
Developed	1.894	0.504
Eastern Europe	−0.661	−1.147
Fossil Fuel Exporters	−1.010	0.604
Asia	−0.220	−0.210
Latin America	0.124	0.005
Africa	0.050	0.008

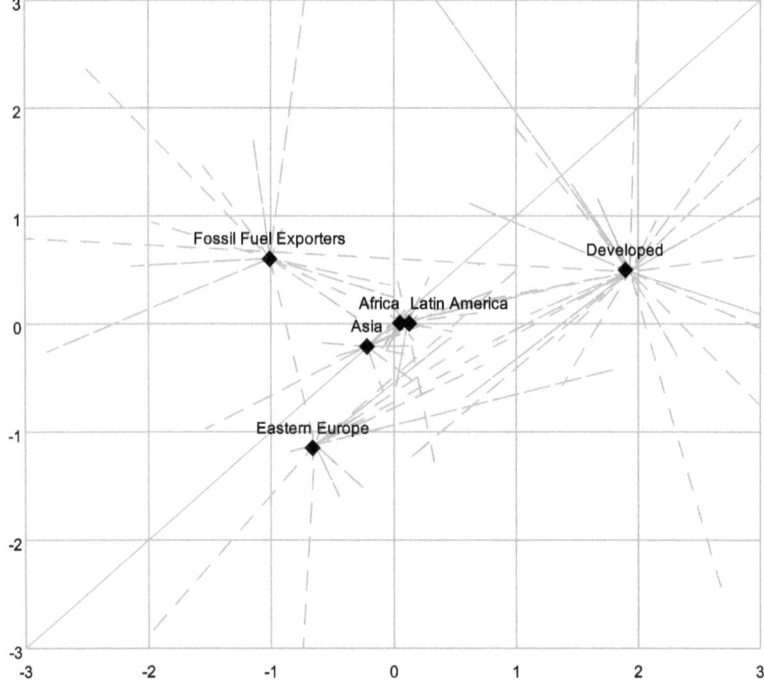

Figure 6.12 Upstream (x-axis) vs downstream carbon trade balance (y-axis) for aggregated world regions (tonCO$_2$ per capita).

In Figure 6.12 we see that Africa, Latin America and Developed Economies have positive UCTB (x-axis). In other words, we can say that the imports of these regions have higher carbon content than their exports. On the other hand, Fossil Fuel Exporters, Eastern Europe and Asia have negative UCTB. This indicates that the exports of these regions have higher carbon content than their imports.

Regarding downstream carbon trade balance (y-axis), Fossil Fuel Exporters, Africa and Developed Economies have positive DCTB. In these regions the payments for exports have higher carbon content than the payments for imports. By contrast, Latin America, Eastern Europe and Asia have negative DCTB, which indicates that the carbon content of their import payments is higher than the carbon content of their export payments.

The carbon trade balance is affected both by the magnitude of the monetary trade balance and by the difference in carbon intensity between import and export flows. More precisely, we can rewrite Eqs. 6.1 and 6.2 as:

$$b_k^U = m_{Wk}^U t_{Wk} - m_k^U t_{kW}$$

$$b_k^D = m_{kW}^D t_{kW} - m_k^D t_{Wk}$$

where b, m and t stand for carbon trade balance, carbon intensity and monetary transaction, U and D stand for upstream or downstream, k is the region of interest and W is the rest of the world. The magnitude of monetary trade flows and carbon intensities of those flows is reported in Table 6.5.

In Table 6.5 we see that Africa and Developed Economies are net importers in economic terms and the upstream intensity of their exports is lower than the upstream intensity of imports. Therefore, it is not surprising to find that they are upstream carbon importers (positive UCTB), as can be seen in Figure 6.12. However, the fact that they are also downstream carbon importers (positive DCTB) is not so obvious, for the same reason that they are net monetary importers. They have positive DCTB because the downstream intensity of their export payments is higher than the downstream intensity of their import payments.

Latin America is a net upstream carbon importer in spite of the surplus in its economic trade balance: this region is a net exporter. Therefore, the goods Latin America produces and exports have an upstream carbon intensity that is substantially lower than that one of its imports. Latin America's negative DCTB is a result of higher intensities at the level of payment for imports, meaning that actually Latin America is sending away more carbon in its payment for imports than it receives in the payment for exports. Thus, Latin America is mainly exporting to less carbon intensive regions.

Table 6.5 Monetary value (10^9 USD 2001) and upstream (U) and downstream (D) carbon intensities and carbon intensities ($kgCO_2$/USD 2001) of exports (index kW or K) and (index Wk or k) imports of aggregated world regions

Region	t_{kW}	t_{Wk}	m_k^U	m_{Wk}^U	m_{kW}^D	m_k^D
Asia	865.74	710.28	1.59	1.01	0.82	1.89
Africa	138.17	152.80	1.20	1.32	1.29	1.16
Developed	1663.42	1969.84	0.71	1.39	1.98	1.43
Eastern Europe	210.26	235.96	2.10	1.13	1.46	2.57
Fossil Fuel E.	1021.05	829.94	1.58	0.96	1.51	1.31
Latin America	351.68	351.49	0.80	0.95	0.84	0.85

Fossil Fuel Exporters is an interesting region since it is the only one that is a net exporter of upstream and importer of downstream embodied emissions (negative UCTB and positive DCTB). The negative UCTB is a result of a surplus in the trade balance added to the fact that the upstream intensity of its exports is higher than the one of its imports. The positive DCTB occurs mainly due to net exports and because the payments for exports have higher downstream intensity than the payments for imports.

Both Eastern Europe and Asia are upstream and downstream carbon importers, but for different reasons. Despite being a net importer, Eastern Europe has a negative UCTB, which is a reflection of the high upstream intensity of exports compared to imports; Asia is a net exporting region whose exports also have higher upstream intensity than its imports. Negative DCTBs are a result of the high downstream intensities of the flows leaving the region (payment for imports), in comparison to the flows entering the region. These economies are very carbon intensive, and export mainly to less carbon intensive ones.

6.5 Total carbon responsibility

So far we have examined consumer and producer carbon responsibility separately. Now we examine how countries and world regions perform regarding total carbon responsibility, which is the arithmetic mean of consumer and producer responsibility, as explained in Chapter 4.

More specifically, we want to know how carbon responsibility deviates from direct emissions, for both GTAP regions and aggregated world regions, in per capita terms. It happens that the difference between carbon responsibility, E_k, and direct emissions, E_k^L, has a simple relation to the carbon trade balances, b_k^U and b_k^D, described in Section 6.4.

We know from the definition of carbon responsibility (Eq. 4.16) that

$$E_k = \frac{1}{2}E_k^C + \frac{1}{2}E_k^P$$

where E_k^C is consumer responsibility and E_k^P is producer responsibility. We also know from the definitions of carbon trade balances (Eqs. 6.1 and 6.2) that:

$$E_k^C = b_k^U + E_k^L$$
$$E_k^D = b_k^D + E_k^L$$

Thus, we can write:

$$E_k - E_k^L = \frac{1}{2}E_k^C + \frac{1}{2}E_k^P - E_k^L$$

$$= \frac{1}{2}\left(b_k^U + E_k^L\right) + \frac{1}{2}\left(b_k^U + E_k^L\right) - E_k^L$$

$$= \frac{1}{2}\left(b_k^U + b_k^D\right)$$

That is, the difference between carbon responsibility and direct emissions, $E_k - E_k^L$, is simply the average of the carbon trade balances. Figures 6.13 and 6.14 show per capita $E_k - E_k^L$ for GTAP regions and for aggregated world regions, respectively. They show per capita UCTB on the x-axis, b_k^U and per capita DCTB on the y-axis, b_k^D (already displayed in Figures 6.11 and 6.12) superimposed on iso-quants of per capita $E_k - E_k^L$. Tables A.8 and 6.4 show, for every GTAP and world region, respectively, absolute

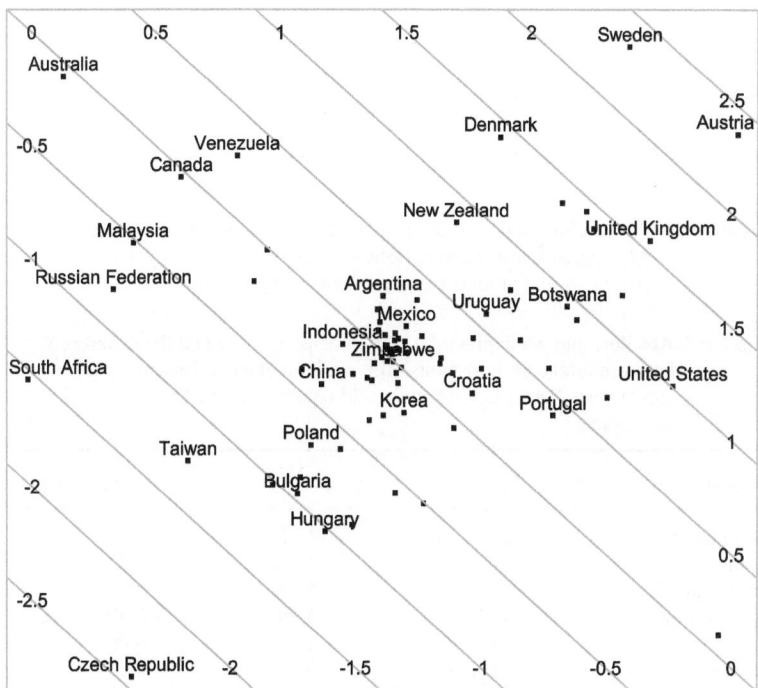

Figure 6.13 Per capita total carbon responsibility minus direct emissions for GTAP regions (tonCO$_2$ per capita) . x and y axis represent consumer and producer responsibility, respectively.

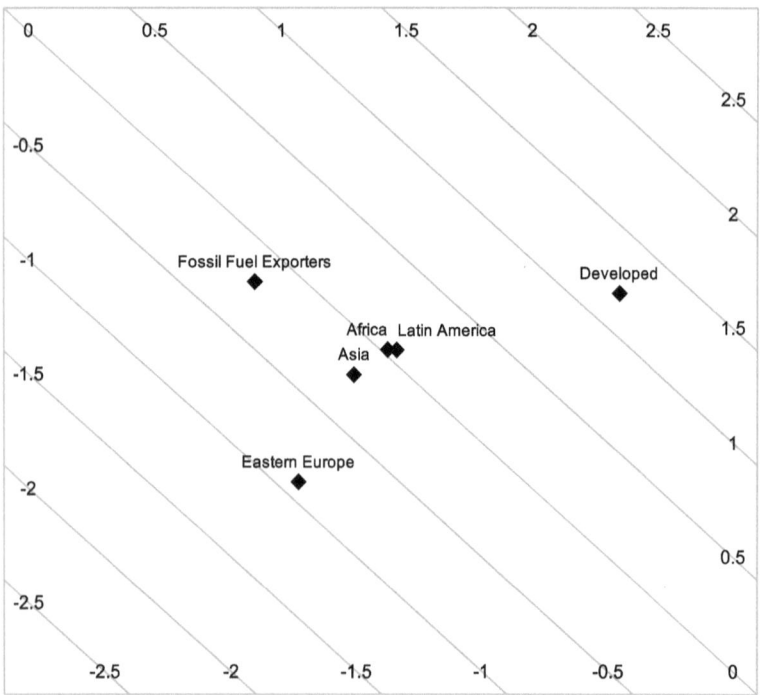

Figure 6.14 Per capita total carbon responsibility minus direct emissions for aggregated world regions (tonCO$_2$ per capita). x and y axis represent consumer and producer responsibility, respectively.

Table 6.6 Absolute and per capita total carbon responsibility and the difference between total carbon responsibility and direct emissions on a per capita basis for six aggregated world regions (MtonCO$_2$ and tonCO$_2$ per capita)

Region	E_k	p.c. E_k	p.c. $(E_k - E_k^L)$
Developed	12221.70	14.75	1.20
Asia	4898.21	1.62	−0.21
Fossil Fuel E.	4704.23	5.71	−0.20
Eastern Europe	1247.79	4.99	−0.90
Latin America	1175.90	2.56	0.06
Africa	502.37	0.68	0.03

and per capita total carbon responsibility, and the difference between total carbon responsibility and direct emissions on a per capita basis.

Countries positioned on the right of the zero iso-quant (representing total carbon responsibility equal to direct emissions) in Figure 6.13 have more total responsibility than direct emissions. This is the case, for example, of Sweden and Austria. Both have about 2.4 tonCO$_2$ per capita more of carbon responsibility than direct emissions. On the other hand, countries positioned on the left of the zero iso-quant have less total responsibility than direct emissions. This is the case, for example, of the Czech Republic, which has almost 2.5 tonCO$_2$ per capita less of carbon responsibility than direct emissions. Australia and Canada are interesting countries because in spite of a large magnitude of UCTB and DCTB values, these cancel out, and their carbon responsibility is close to their direct emissions.

Looking at Figure 6.14, we find that for Developed Economies, Africa and Latin America, total carbon responsibility is higher than direct emissions, while for Fossil Fuel Exporters, Asia and Eastern Europe the opposite is true.

7   Discussion

7.1 Summary

In this book we addressed the problems of defining and measuring carbon responsibility and embodied emissions.

In Chapter 2 we reviewed the mathematical techniques required to compute total embodied carbon emissions, using input–output (IO) analysis. Total upstream embodied carbon emissions (TUECEF) and total downstream embodied carbon emissions (TDECEF) of a given economic flow are numbers that indicate the amount of direct and indirect carbon emissions that take place in order to, respectively, generate the economic flow or the monetary payment thereof. That is, TUECEF and TDECEF respectively account for the upstream and downstream emissions throughout the life cycle of an economic flow.

We then addressed the problem of defining an indicator applied to economic agents that takes into account (upstream and downstream) indirect carbon emissions. In Chapter 3 we reviewed a number of existing indicators and in Chapter 4 presented carbon responsibility. Total carbon responsibility is an indicator that satisfies six important properties: normalization (the carbon responsibility of the world matches total direct emissions); scale invariance (economic agents can be disaggregated without affecting the distribution of responsibilities); monotonicity (responsibility is monotonic in direct emissions); accounting of indirect effects (carbon responsibility is a function of embodied emissions); economic causality (a rule to allocate embodied emissions in the case of multiple inputs or outputs) and symmetry (carbon responsibility remains unchanged if upstream and downstream embodied emissions are interchanged as arguments).

Total carbon responsiblity of an economic agent is the average between that agent's consumer and producer carbon responsibilities, where consumer carbon responsibility is the sum of the TUECEF of final demand, and producer carbon responsibility is the sum of the TDECEF of primary inputs.

After defining the quantities of total embodied carbon emissions and carbon responsibilities, we presented a multi-regional IO model (MRIO), in order to empirically measure those quantities. We used the GTAP 6 database, which partitions the world into 87 regions, with 57 economic sectors each, and reports data for the year 2001. In Chapter 5 we reported the procedure for the construction of the MRIO model, and in Chapter 6 we reported the results.

We performed a statistical analysis on the fraction of total embodied emissions that result from domestic and international indirect effects, and on the uncertainty from the estimation of international trade flows, which affects MRIO models but not domestic IO models. We concluded that international indirect effects are on average less important than domestic indirect effects but still significant, and that the magnitude of the uncertainty is much smaller than that of international indirect effects, showing that a full multi-regional IO model can and should be used to account for total embodied emissions.

We observed that direct emissions, consumer and producer responsibility all increase with GDP, having positive elasticities with magnitude smaller than one, and consumer responsibility is the type of emissions with the highest elasticity to income.

Next, we compared responsibilities against direct emissions. We observed that small open economies have carbon responsibilities that differ substantially from total direct emissions. About 90% of world direct emissions and responsibilities (consumer, producer and total responsibility) result from the economic activity of Asia, Developed Economies and Fossil Fuel Exporters. Developed Economies have both consumer and producer responsibility higher than direct emissions; Asian and Eastern European countries have both responsibilities lower than direct emissions; Fossil Fuel Exporters have consumer responsibility lower and producer responsibility higher than direct emissions; Latin America and Africa do not display noticeable differences between either of those quantities. The explanation for these patterns lies in the carbon content of bilateral trade, because each region displays different magnitudes of monetary imports and exports, as well as different carbon intensities. Differences from total carbon responsibility to direct emissions are more pronounced at the country level.

7.2 Open questions

The work we reported uses multi-regional IO analysis to quantify upstream and downstream embodied carbon emissions of economic flows, and carbon responsibilities of economic agents. The use of these quantities for practical purposes requires answering open empirical and methodological questions.

On the empirical side, better IO databases are necessary, ones that present detailed and recently updated national statistical data in a unified framework. The available databases have a small resolution and are either incomplete (meaning they do not cover the whole world, as the OECD or EUROSTAT tables) or present data of weak quality (as the GTAP database). There is an ongoing EU project, called EXIOPOL (Tukker *et al.* 2009) that attempts to solve this problem, but substantial work on this topic is urgently needed.

On the methodological side, a standard and thorough method of error accounting is also necessary. Both LCA and IO techniques require the use of empirical data that is inaccurate, and the uncertainty of either source data or results is seldom reported. We believe that carbon labelling can only be credible if errors are systematically tracked, because otherwise society at large will not trust the numbers. There are isolated studies that report uncertainties (Lenzen 2001, Wiedmann *et al.* 2008a), but the accounting of uncertainties should become the rule, rather than the exception.

7.3 Embodied emissions

7.3.1 Carbon labelling

We believe embodied carbon emissions have the potential to play a role in climate change policy.

In a first stage, we believe labelling the TUECEF of goods and services would allow economic agents to compare products on the basis of their environmental performance and to make informed choices, balancing their economic preferences (price) and their environmental preferences (embodied emissions).

Such carbon labelling would be quantitative, as opposed to qualitative, such as 'organic food', which does not provide accurate information about the impacts of the product (Economist 2008). We consider that qualitative labels are potentially misleading, as in the example of 'food miles'. In the UK retail sector in recent years, a label of the distance from food production to store was used, because supposedly there is a correlation between carbon emissions and distance, but that correlation is in fact weak (Muller 2007). Of course, qualitative labels also have merits: they cause less 'cognitive overload' on the consumer; they easily integrate different dimensions; and they may be more adequate when there are high levels of uncertainty.

Thus, in the near future we would like to see the increased use of a quantitative label of total carbon embodied emissions, following a clear measurement methodology and thorough error accounting. Some steps are being taken in this direction, especially in the area of final consumption products. Many life cycle assessment (LCA) studies (EPLCA 2008) of specific products have been published (Tukker and Jansen 2006), and the British Standards are advancing towards the publication of a standard for the quantification of the TUECEF of products (BSI 2008).

The British standard uses LCA techniques only, but we believe that in the near future, another standard will be adopted that combines both LCA and MRIO analysis, in order to report more accurate results. We believe that the widespread use of carbon labelling, using IO techniques, can only take off if both empirical and methodological problems referred to in Section 7.2 are solved.

In spite of the remaining difficulties, we believe the upstream carbon labelling of final consumption products is slowly becoming a reality.

The labelling of TUECEF of final consumption products generates a cascade effect: for a retailer to label his products, he must know the embodied emissions of his own inputs, and thus, upstream carbon labelling has the potential to spread upstream throughout the production chain, eventually raising environmental awareness among firms, in their role as intermediate consumers.

7.3.2 Downstream embodied emissions

In this book we have systematically drawn attention to downstream embodied emissions, which, with few exceptions (Gallego and Lenzen 2005), have been ignored in the literature.

We believe that in the same way that economic agents can decide on the environmental impact of the goods and services they buy (a consumer role), they can also decide on the environmental impact of the goods and services they sell (producer role).

All firms sell goods or services, either intermediate or final, and households sell their labour and capital to firms. This represents a huge number of decisions made by individual economic agents. Agents could express their environmental preferences in these decisions if they were informed of the environmental consequences of their actions: that is, if they had access to the downstream carbon labelling of the recipients of their goods or services.

We may think that a seller does not have options (Cerin 2006): e.g., for an unskilled worker the option can be to work at a smoky factory or to stay unemployed (which is not an attractive option); e.g., once a consumer good is placed on a grocery shelf, the selling firm is blind to the future use of the product. However, in some circumstances, sellers have the option of making well informed choices.

The Equator Principles (http://www.equator-principles.com/) are a set of principles that financial institutions can voluntarily adopt 'to ensure that the projects [they] finance are developed in a manner that is socially responsible and reflect[s] sound environmental management practice' (from the Preamble).

Likewise, the Association of British Insurers (ABI 2004) reports that '[i]n 2001 […] a third of [financial] analysts said social and environmental policies were important in helping them assess companies' and that accumulated evidence in the past three decades lends overwhelming weight to the view that 'investors can enhance risk/return performance through a better understanding of the social and environmental risks companies face and their skills in managing their risks' (ABI 2004: 5).

The Equator Principles and the Association of British Insurers espouse the concept of accounting for downstream indirect effects, because the financial institution (which sells a financial product) is concerned about the environmental impacts occurring downstream (that is, resulting from the actions of the buyer of that product).

If we change the focus from environmental to social issues, responsibility for downstream effects becomes much more familiar: some people oppose the manufacture and trade of weapons, tobacco or illegal drugs. Such products have adverse social effects (war, addiction, crime) downstream along the economic process. Hence, if an individual refuses to engage in the manufacture or trade of such products, and in order not to receive 'dirty money', that individual is acknowledging the principle of accounting for downstream indirect effects.

The accounting of TDECEF of goods and services is not something radically new. It is the operationalization in a quantitative labelling scheme of ideas that already exist but only in a qualitative manner (Heal 2005).

7.4 Carbon responsibility

7.4.1 Environmental agreements

Climate change is a global environmental problem, as the GHG emissions of one particular economic agent are going to affect the welfare of all other economic agents in the world. Under the auspices of the United Nations, representatives of all nations gathered to design the Kyoto Protocol, an international environmental agreement whose purpose is to curb global greenhouse gas emissions.

Other international environmental agreements were designed earlier, and so we can capitalize on their experience to analyse how the potential use of carbon responsibility as an environmental indicator can affect the outcome of such an agreement.

The purpose of an international environmental agreement is to bring as many interested parties (in this case, GHG emitters) into the agreement as possible, and to allocate abatement effort among the participants, maximizing the welfare of all participants (an efficiency requirement), while taking into account the distribution of costs and benefits among agents (an equity requirement) (Barrett 1994, Carraro and Siniscalco 2003, Hoel and Schneider 1997).

The balancing of efficiency and equity requirements in an international environmental agreement is non-trivial, given the heterogeneity of real world agents. Among others, Chichilnisky and Heal (1994) showed that, in the presence of a public good, it is impossible to separate efficiency and equity.

Equity preferences and the perception that the agreement is fair are important components for the success of an agreement. Lange (2006) writes that '[c]hances of cooperation are only increased if (…) free-riding incentives are limited. This is implicitly the case if countries are inequality-averse with respect to abatement effort.' In other words, participants are more likely to cooperate if they perceive that the allocation of abatement effort is equitable.

Along the same line of thought, Albin (2003) writes that fairness is required in all steps of a negotiation process: to bring all parties to the negotiation table, to evaluate alternative proposals, to overcome conflicting interests and reach agreement, to legitimise the outcome before the constituency, to secure implementation and compliance and to establish and maintain long-term cooperation. Perception of unfairness at any of these steps causes mistrust and reduces the likelihood of cooperation. According to this author, there are different principles of fairness, to be applied at the different negotiation steps, but they must all fall under the same overarching concept of justice: fairness is the balanced settlement of conflicting interests.

Given the important role that fairness and equity preferences play in the success of an international environmental agreement, it is interesting to note that the choice of the indicator used in the negotiation has not yet been fully explored (Hoekstra and Janssen 2006, Peters and Hertwich 2008b). For practical purposes, all international environmental agreements in the real world so far use direct emissions as environmental indicators.

As we saw in Chapter 6, the choice of indicator affects the apparent contribution of a country to global emissions. We believe that the use of carbon responsibility instead of direct carbon emissions as environmental indicator could have a positive role in the success of environmental negotiations. For example, instead of committing to reduce its direct carbon emissions by x%, a country would commit to reduce its carbon responsibility by y%. We believe carbon responsibility displays a country's contribution to global emissions that is more in line with the intuitive perception of that contribution, and its use would thus raise trust in the negotiation process and in the fairness of an international environmental agreement.

In the work reported in this book, we have considered only current carbon emissions, i.e. the carbon responsibility of a country is a function of embodied carbon emissions generated during a short time span (e.g., one year). However, the damage resulting from climate change and the historical contribution of carbon emissions are unevenly distributed across the globe, with tropical poor countries suffering most damage and temperate rich countries contributing most to historical emissions (Srinivasan *et al.* 2008). This observation has led some authors to propose the use of historical direct carbon emissions in the context of a climate change agreement (Panayotou *et al.* 2002). With appropriate data, not only current but also historical carbon responsibilities could be computed. However, the adding of historical emissions is a non-trivial question, due to the non-linear effect of CO_2 emissions on climate.

7.4.2 Environmental policy

In Section 7.3 we mentioned the potential use of embodied emissions in voluntary carbon abatement: if carbon labelling is implemented, economic agents can choose economic inputs and outputs on the basis of their carbon intensities.

However, the largest contribution to the current effort to curb carbon emissions comes not from voluntary abatement but from government-sponsored carbon abatement policies. Such policies involve one of three mechanisms: standards, permits, or taxes/subsidies (Hanley *et al.* 2007).

Standards are detailed rules with which economic agents must comply. Standards play a role in climate change policy (in EU countries, for example, industrial units and households must conform to energy efficiency standards); but the environmental economics and policy literature pays more attention to market based mechanisms: permits (quantity control) and taxes/subsidies (price control) (Pizer 2002), which we now address.

Tradeable carbon permits, as currently used in the EU to control carbon emissions from large point emitters (Grubb and Neuhoff 2006), are allocated to economic agents, which can either use the permits or trade them in a market. The system of tradeable permits works by setting a total number of permits which is lower than the current level of emissions. The trading of permits between firms guarantees that abatement is carried out by those firms for whom it is least costly (Ellerman 2005). However, the initial allocation of tradeable permits is a delicate problem (Jesper and Rasmussen 2000).

Carbon taxes (applied to goods of high carbon intensity) and subsidies (applied to goods of low carbon intensity) are another policy tool currently used to stimulate carbon abatement (Bazin *et al.* 2004). Carbon taxes are easier to monitor than carbon permits but they can be politically less acceptable, and they pose the equity problem of distributing the tax revenues (Ekins and Barker 2001).

There is debate concerning the merits of taxes and permits as effective policy tools, especially in the context of free trade and heterogeneous world regions (Copeland and Taylor 2005, Ekins and Barker 2001, Mandell 2008, Pezzey 2003). With the exceptions of Hoekstra and Janssen (2006) and Peters and Hertwich (2008b), all the studies of climate change policy we are aware of use direct carbon emissions as the environmental indicator.

However, it is possible to create a carbon tax or to allocate carbon permits using carbon responsibility as an environmental indicator. Carbon responsibility, compared to direct emissions, has the disadvantage of being more difficult to quantify and to implement, but it has the advantage (from the point of view of equity) of being inclusive: most direct emissions originate in a few economic sectors (energy and heavy industry), but embodied emissions trickle down through the economic network, and virtually all economic agents have some carbon responsibility.

Using carbon responsibility as an environmental indicator, the burden of carbon abatement is distributed among all economic agents. Furthermore, carbon responsibility has the potential of being more easily accepted by economic agents, given the discussion on equity preferences and fairness developed in Section 7.4.1.

Of course, it is an open issue whether the advantages of using carbon responsibility outweigh the disadvantages. We do not blindly advocate the use of carbon responsibility. Instead, we believe that it is worth exploring the efficiency and equity implications of carbon responsibility for environmental policy.

Appendix A

A.1 List of symbols

The most frequently used symbols in the book are summarized in Table A.1.

Table A.1 List of frequent symbols

Symbol	Definition
e	Carbon emission of sector or flow
m	Carbon intensity of sector or flow
E	Carbon responsibility of a region
b	Carbon trade balance of a region
t	Economic flow (or transaction)
ex	Economic flow (export)
im	Economic flow (import)
N	Number
R	Set of regions
S	Set of sectors
F	Set of flows

Subscript	Definition
i,j,l	Indices denoting sector (ij means from i to j)
k,k',k"	Indices denoting region
W	World (the set of all regions)
R	Region
S	Sector
*	Sum over all sectors

Table A.1 Cont'd

Superscript	Definition
L	Direct (or local)
U	Upstream embodied
D	Downstream embodied
C	Consumer
P	Producer
i	International
d	Domestic
a, b	Indices denoting region
*	Sum over all regions

A.2 GTAP sector and region codes

Table A.2 lists the GTAP sector codes. The GTAP lists agricultural sectors first, then industrial sectors and finally service sectors. In the GTAP nomenclature 'nec' means 'not elsewhere classified'.

Table A.2 List of GTAP sector codes (nec = not elsewhere classified)

$N°$	Code	Sector
1	pdr	Paddy rice
2	wht	Wheat
3	gro	Cereal grains (nec)
4	v_f	Vegetables, fruit, nuts
5	osd	Oil seeds
6	c_b	Sugar cane, sugar beet
7	pfb	Plant-based fibres
8	ocr	Crops (nec)
9	ctl	Cattle, sheep, goats, horses
10	oap	Animal products (nec)
11	rmk	Raw milk
12	wol	Wool, silk-worm cocoons
13	frs	Forestry
14	fsh	Fishing
15	coa	Coal
16	oil	Oil
17	gas	Gas

continued

Table A.2 Cont'd

N°	Code	Sector
18	omn	Minerals (nec)
19	cmt	Meat: cattle, sheep, goats, horses
20	omt	Meat products (nec)
21	vol	Vegetable oils and fats
22	mil	Dairy products
23	pcr	Processed rice
24	sgr	Sugar
25	ofd	Food products (nec)
26	b_t	Beverages and tobacco products
27	tex	Textiles
28	wap	Wearing apparel
29	lea	Leather products
30	lum	Wood products
31	ppp	Paper products, publishing
32	p_c	Petroleum, coal products
33	crp	Chemical, rubber, plastic products
34	nmm	Mineral products (nec)
35	i_s	Ferrous metals
36	nfm	Metals (nec)
37	fmp	Metal products
38	mvh	Motor vehicles and parts
39	otn	Transport equipment (nec)
40	ele	Electronic equipment
41	ome	Machinery and equipment (nec)
42	omf	Manufactures (nec)
43	ely	Electricity
44	gdt	Gas manufacture, distribution
45	wtr	Water
46	cns	Construction
47	trd	Trade
48	otp	Transport (nec)
49	wtp	Sea transport
50	atp	Air transport
51	cmn	Communication
52	ofi	Financial services (nec)
53	isr	Insurance
54	obs	Business services (nec)
55	ros	Recreation and other services
56	osg	Public administration, defence, health, education
57	dwe	Dwellings

Table A.3 lists the GTAP region codes, and Table A.4 lists the countries that form GTAP's composite regions. GTAP regions are ordered first by continent (Oceania, Asia, Americas, Europe and Asia) and within continent in alphabetical order. In the GTAP nomenclature, a region code whose first letter is 'x' is a composite region.

Table A.3 List of GTAP region codes

	Code	Country		Code	Country
1	aus	Australia	38	bel	Belgium
2	nzl	New Zealand	39	dnk	Denmark
3	xoc	Rest of Oceania	40	fin	Finland
4	chn	China	41	fra	France
5	hkg	Hong Kong	42	deu	Germany
6	jpn	Japan	43	gbr	United Kingdom
7	kor	Korea	44	grc	Greece
8	twn	Taiwan	45	irl	Ireland
9	xea	Rest of East Asia	46	ita	Italy
10	idn	Indonesia	47	lux	Luxembourg
11	mys	Malaysia	48	nld	The Netherlands
12	phl	Philippines	49	prt	Portugal
13	sgp	Singapore	50	esp	Spain
14	tha	Thailand	51	swe	Sweden
15	vnm	Vietnam	52	che	Switzerland
16	xse	Rest of South-east Asia	53	xef	Rest of EFTA
17	bgd	Bangladesh	54	xer	Rest of Europe
18	ind	India	55	alb	Albania
19	lka	Sri Lanka	56	bgr	Bulgaria
20	xsa	Rest of South Asia	57	hrv	Croatia
21	can	Canada	58	cyp	Cyprus
22	usa	United States	59	cze	Czech Republic
23	mex	Mexico	60	hun	Hungary
24	xna	Rest of North America	61	mlt	Malta
25	col	Colombia	62	pol	Poland
26	per	Peru	63	rom	Romania
27	ven	Venezuela	64	svk	Slovakia
28	xap	Rest of Andean Pact	65	svn	Slovenia
29	arg	Argentina	66	est	Estonia
30	bra	Brazil	67	lva	Latvia
31	chl	Chile	68	ltu	Lithuania
32	ury	Uruguay	69	rus	Russian Federation
33	xsm	Rest of South America	70	xsu	Rest of former Soviet Union
34	xca	Central America	71	tur	Turkey
35	xfa	Rest of FTAA	72	xme	Rest of Middle East
36	xcb	Rest of the Caribbean	73	mar	Morocco
37	aut	Austria	74	tun	Tunisia

continued

Table A.3 Cont'd

	Code	Country		Code	Country
75	xnf	Rest of North Africa	82	zmb	Zambia
76	bwa	Botswana	83	zwe	Zimbabwe
77	zaf	South Africa	84	xsd	Rest of SADC
78	xsc	Rest of South African CU	85	mdg	Madagascar
79	mwi	Malawi	86	uga	Uganda
80	moz	Mozambique	87	xss	Rest of Sub-Saharan Africa
81	tza	Tanzania			

Table A.4 List of countries that form GTAP's composite regions

Region	Countries
xoc	American Samoa, Cook Islands, Fiji, French Polynesia, Guam, Kiribati, Marshall Islands, Federated States of Micronesia, Nauru, New Caledonia, Niue, Norfolk Island, Northern Mariana Islands, Palau, Papua New Guinea, Samoa, Solomon Islands, Tokelau, Tonga, Tuvalu, Vanuatu, Wallis and Futuna
xea	Democratic People's Republic of Korea, Macau, Mongolia
xse	Brunei Darussalam, Cambodia, Lao People's Democratic Republic, Myanmar, Timor Leste
xsa	Afghanistan, Bhutan, Maldives, Nepal, Pakistan
xna	Bermuda, Greenland, Saint Pierre and Miquelon
xap	Bolivia, Ecuador
xsm	Falkland Islands (Malvinas), French Guiana, Guyana, Paraguay, Suriname
xca	Belize, Costa Rica, El Salvador, Guatemala, Honduras, Nicaragua, Panama
xfa	Antigua and Barbuda, Bahamas, Barbados, Dominica, Dominican Republic, Grenada, Haiti, Jamaica, Puerto Rico, Saint Kitts and Nevis, Saint Lucia, Saint Vincent and the Grenadines, Trinidad and Tobago, US Virgin Islands
xcb	Anguilla, Aruba, Cayman Islands, Cuba, Guadeloupe, Martinique, Montserrat, Netherlands Antilles, Turks and Caicos, British Virgin Islands
xef	Iceland, Liechtenstein, Norway
xer	Andorra, Bosnia and Herzegovina, Faroe Islands, Gibraltar, the former Yugoslav Republic of Macedonia, Monaco, San Marino, Serbia and Montenegro
xsu	Armenia, Azerbaijan, Belarus, Georgia, Kazakhstan, Kyrgyzstan, Republic of Moldova, Tajikistan, Turkmenistan, Ukraine, Uzbekistan

Table A.4 Cont'd

Region	Countries
xme	Bahrain, Islamic Republic of Iran, Iraq, Israel, Jordan, Kuwait, Lebanon, Oman, Occupied Palestinian Territory, Qatar, Saudi Arabia, Syrian Arab Republic, United Arab Emirates, Yemen
xnf	Algeria, Egypt, Libyan Arab Jamahiriya
xsc	Lesotho, Namibia, Swaziland
xsd	Angola, Congo, Mauritius, Seychelles
xss	Benin, Burkina Faso, Burundi, Cameroon, Cape Verde, Central African Republic, Chad, Comoros, Congo, Cote d'Ivoire, Djibouti, Equatorial Guinea, Eritrea, Ethiopia, Gabon, Gambia, Ghana, Guinea, Guinea-Bissau, Kenya, Liberia, Mali, Mauritania, Mayotte, Niger, Nigeria, Reunion, Rwanda, Saint Helena, Sao Tome and Principe, Senegal, Sierra Leone, Somalia, Sudan, Togo

Table A.5 lists the GTAP regions that form the aggregated world regions.

Table A.5 List of GTAP regions in each of six world regions

World region	GTAP regions
Asia (A)	China, India, Korea, Taiwan, Thailand, rest of South Asia, rest of East Asia, Philippines, Vietnam, Bangladesh, rest of South-east Asia, Sri Lanka
Africa (AF)	Turkey, rest of Sub-Saharan Africa, Morocco, Tunisia, rest of SADC, Zimbabwe, Tanzania, Rest of South African CU, Uganda, Botswana, Madagascar, Zambia, Mozambique, Malawi
Developed (D)	United States, Japan, Germany, United Kingdom, France, Italy, Spain, The Netherlands, Belgium, Greece, Finland, Denmark, rest of Europe, Portugal, Austria, Sweden, Hong Kong, Switzerland, Singapore, Ireland, New Zealand, Luxembourg, Cyprus, rest of Oceania, Malta, rest of North America
Fossil Fuel Exporter (F)	Rest of EFTA, Russian Federation, rest of Middle East, Canada, Australia, South Africa, Indonesia, rest of North Africa, Venezuela, Malaysia, Colombia
Latin America (L)	Mexico, Brazil, Argentina, rest of FTAA, Chile, rest of the Caribbean, Central America, rest of Andean Pact, Peru, rest of South America, Uruguay
Eastern Europe (E)	Rest of Former Soviet Union, Poland, Czech Republic, Romania, Hungary, Bulgaria, Slovakia, Croatia, Slovenia, Estonia, Lithuania, Latvia, Albania

A.3 Carbon responsibilities of GTAP regions

Table A.6 reports absolute direct emissions, consumer and producer carbon responsibilities (E_k^L, E_k^U and E_k^D) and error margins for each GTAP region k, for the year 2001 (MtonCO$_2$ or 10^9 kgCO$_2$).

Table A.6 Absolute direct emissions, consumer and producer responsibility, with error margins, of all GTAP regions for the year 2001 (MtonCO$_2$), ordered by decreasing absolute direct emissions

Region	Code	E_k^L	E_k^U	ϵ	E_k^D	ϵ
United States	usa	6006.93	6654.50	15.81	5918.22	5.96
China	chn	3289.16	2690.60	10.02	2905.56	5.41
Russian Federation	rus	1502.84	1190.46	1.21	1580.83	9.85
Japan	jpn	1290.96	1506.89	8.00	1425.98	2.07
Rest of Middle East	xme	1199.90	1043.28	10.64	1352.83	7.50
India	ind	1024.77	971.22	3.01	901.14	0.67
Germany	deu	892.20	1011.18	10.78	998.57	6.63
Rest of Former Soviet Union	xsu	713.88	595.70	2.41	550.44	2.32
United Kingdom	gbr	618.59	746.12	3.21	675.50	2.87
Canada	can	547.66	497.81	4.85	595.26	1.75
France	fra	509.92	607.78	5.64	582.66	1.66
Italy	ita	475.09	539.24	16.14	480.30	1.94
Korea	kor	397.72	406.40	3.75	371.46	0.48
Mexico	mex	389.92	410.24	2.94	411.31	2.18
Australia	aus	351.55	302.10	0.64	398.32	3.03
South Africa	zaf	323.67	200.86	0.27	312.34	0.95
Brazil	bra	320.98	321.94	3.93	309.66	0.96
Poland	pol	309.82	288.32	4.72	277.39	0.45
Spain	esp	305.71	337.40	1.73	299.26	2.32
Indonesia	idn	305.37	240.68	0.99	316.85	2.08
Rest of North Africa	xnf	248.56	245.75	2.02	284.79	5.25
Taiwan	twn	247.91	213.60	2.98	226.21	0.51
The Netherlands	nld	204.52	225.97	2.05	201.97	1.04
Turkey	tur	196.29	189.06	4.43	179.95	0.43
Thailand	tha	178.68	139.34	1.14	167.92	0.68
Rest of Sub-Saharan Africa	xss	169.49	190.34	4.56	187.96	2.64
Venezuela	ven	155.80	127.59	0.61	198.05	4.57
Belgium	bel	126.43	142.55	2.63	129.18	0.60
Argentina	arg	120.41	121.02	1.09	138.30	2.61
Czech Republic	cze	117.37	96.93	0.63	87.96	0.27
Malaysia	mys	116.77	69.88	0.85	139.09	1.43
Rest of South Asia	xsa	108.80	118.59	0.51	89.64	0.21
Rest of FTAA	xfa	106.58	107.05	0.99	91.65	0.66
Rest of East Asia	xea	102.02	99.67	1.08	86.21	0.20

Table A.6 Cont'd

Region	Code	E_k^L	E_k^U	ϵ	E_k^D	ϵ
Greece	grc	101.36	107.51	1.16	94.11	0.30
Finland	fin	94.86	77.26	0.35	99.10	1.20
Romania	rom	93.64	86.45	0.83	74.17	0.57
Hungary	hun	82.16	77.74	0.66	66.20	0.52
Philippines	phl	75.98	85.66	2.07	59.77	0.25
Denmark	dnk	74.77	79.87	0.97	84.77	0.32
Rest of Europe	xer	74.07	75.97	0.59	52.71	0.10
Portugal	prt	70.01	83.75	0.68	64.28	0.50
Austria	aut	67.22	90.27	1.20	82.59	0.50
Colombia	col	62.13	61.79	0.50	72.72	0.57
Vietnam	vnm	59.75	70.53	3.17	36.55	1.37
Sweden	swe	59.71	77.34	1.43	83.41	0.89
Rest of EFTA	xef	57.35	59.30	1.65	117.88	3.93
Hong Kong	hkg	54.31	142.72	4.16	106.02	2.61
Chile	chl	51.95	51.66	0.46	54.63	0.15
Switzerland	che	51.33	102.59	1.55	96.25	1.17
Singapore	sgp	48.93	74.02	2.50	53.32	0.69
Rest of the Caribbean	xcb	44.80	49.05	0.89	27.90	0.51
Ireland	irl	44.34	43.53	0.53	58.99	0.38
Bulgaria	bgr	41.01	35.64	1.12	30.76	0.49
Central America	xca	40.26	57.90	1.14	37.60	0.29
Morocco	mar	38.23	42.97	0.62	33.65	0.25
Slovakia	svk	37.06	35.82	1.24	28.72	0.35
Bangladesh	bgd	36.72	48.33	1.07	31.60	0.18
Rest of Andean Pact	xap	33.79	36.89	0.23	33.78	0.31
New Zealand	nzl	27.35	29.67	0.28	31.68	0.14
Peru	per	27.16	32.15	0.64	29.17	0.09
Tunisia	tun	24.50	23.80	0.36	21.90	0.40
Croatia	hrv	20.08	23.31	0.61	18.40	0.18
Rest of SADC	xsd	19.69	21.88	0.60	28.71	1.25
Slovenia	svn	18.43	19.37	0.09	18.20	0.14
Estonia	est	17.70	16.57	0.18	11.97	0.09
Rest of Southeast Asia	xse	16.04	23.64	0.48	25.97	0.29
Zimbabwe	zwe	12.93	14.13	0.13	11.69	0.03
Sri Lanka	lka	11.58	15.61	0.38	11.21	0.19
Lithuania	ltu	11.37	15.18	0.87	13.32	0.61
Luxembourg	lux	11.07	14.96	0.23	10.17	0.06
Cyprus	cyp	8.21	10.26	0.14	6.30	0.05
Latvia	lva	7.62	11.93	0.74	6.62	0.07
Rest of Oceania	xoc	6.23	8.87	0.20	7.23	0.05
Rest of South America	xsm	5.27	7.31	0.05	8.40	0.25

continued

Table A.6 Cont'd

Region	Code	E_k^L	E_k^U	ϵ	E_k^D	ϵ
Uruguay	ury	5.14	7.96	0.14	6.21	0.12
Albania	alb	3.60	5.58	0.17	2.91	0.01
Tanzania	tza	3.60	6.32	0.18	3.63	0.03
Rest of South African CU	xsc	3.56	7.29	0.39	4.10	0.04
Uganda	uga	3.09	4.73	0.06	2.55	0.02
Botswana	bwa	2.64	5.06	0.27	3.27	0.08
Malta	mlt	2.45	3.97	0.14	1.77	0.03
Madagascar	mdg	2.34	3.50	0.11	2.08	0.01
Zambia	zmb	2.04	3.47	0.12	3.08	0.30
Rest of North America	xna	1.90	3.59	0.07	1.26	0.01
Mozambique	moz	1.60	3.55	0.13	3.13	0.67
Malawi	mwi	0.99	1.91	0.05	1.04	0.03

Table A.7 reports the per capita direct emissions, consumer and producer carbon responsibilities (p.c. E_k^L, p.c. E_k^U and p.c. E_k^D) and error margins for every GTAP region k. (tonCO$_2$ per capita.)

Table A.7 Per capita direct emissions, consumer and producer responsibility, with error margins, of all GTAP regions (tonCO$_2$ p.c.), ordered by decreasing per capita direct emissions

Region	Code	p.c. E_k^L	p.c. E_k^U	ϵ	p.c. E_k^D	ϵ
Luxembourg	lux	25.10	33.92	0.53	23.06	0.14
United States	usa	21.65	23.98	0.06	21.33	0.02
Finland	fin	18.28	14.89	0.07	19.09	0.23
Australia	aus	18.10	15.55	0.03	20.50	0.16
Canada	can	17.53	15.94	0.16	19.06	0.06
Rest of North America	xna	15.07	28.48	0.54	9.97	0.04
Singapore	sgp	14.69	22.23	0.75	16.01	0.21
Denmark	dnk	14.00	14.96	0.18	15.88	0.06
The Netherlands	nld	12.81	14.15	0.13	12.65	0.07
Estonia	est	12.40	11.60	0.12	8.38	0.06
Belgium	bel	12.29	13.86	0.26	12.56	0.06
Rest of EFTA	xef	11.88	12.28	0.34	24.41	0.81
Ireland	irl	11.75	11.53	0.14	15.63	0.10
Czech Republic	cze	11.48	9.48	0.06	8.60	0.03
Taiwan	twn	11.12	9.58	0.13	10.15	0.02
Germany	deu	10.88	12.33	0.13	12.18	0.08

Table A.7 Cont'd

Region	Code	p.c. E_k^L	p.c. E_k^U	ϵ	p.c. E_k^D	ϵ
Cyprus	cyp	10.79	13.49	0.18	8.28	0.07
United Kingdom	gbr	10.44	12.59	0.05	11.40	0.05
Russian Federation	rus	10.31	8.17	0.01	10.85	0.07
Japan	jpn	10.18	11.88	0.06	11.25	0.02
Greece	grc	9.59	10.18	0.11	8.91	0.03
Slovenia	svn	9.25	9.73	0.05	9.14	0.07
France	fra	8.57	10.21	0.09	9.79	0.03
Korea	kor	8.36	8.54	0.08	7.81	0.01
Austria	aut	8.30	11.14	0.15	10.20	0.06
Italy	ita	8.26	9.37	0.28	8.35	0.03
Hungary	hun	8.22	7.78	0.07	6.63	0.05
Poland	pol	8.00	7.44	0.12	7.16	0.01
Spain	esp	7.75	8.56	0.04	7.59	0.06
Hong Kong	hkg	7.58	19.92	0.58	14.80	0.36
South Africa	zaf	7.46	4.63	0.01	7.20	0.02
Switzerland	che	7.17	14.32	0.22	13.44	0.16
New Zealand	nzl	7.11	7.71	0.07	8.23	0.04
Portugal	prt	7.00	8.37	0.07	6.43	0.05
Rest of Middle East	xme	6.91	6.01	0.06	7.79	0.04
Slovakia	svk	6.84	6.61	0.23	5.30	0.06
Sweden	swe	6.73	8.72	0.16	9.40	0.10
Venezuela	ven	6.31	5.17	0.02	8.03	0.19
Malta	mlt	6.21	10.05	0.35	4.48	0.07
Rest of Former Soviet Union	xsu	5.18	4.33	0.02	4.00	0.02
Bulgaria	bgr	5.05	4.39	0.14	3.79	0.06
Malaysia	mys	4.93	2.95	0.04	5.88	0.06
Croatia	hrv	4.54	5.27	0.14	4.16	0.04
Rest of Europe	xer	4.37	4.48	0.03	3.11	0.01
Romania	rom	4.19	3.87	0.04	3.32	0.03
Rest of FTAA	xfa	4.11	4.13	0.04	3.54	0.03
Rest of East Asia	xea	3.99	3.90	0.04	3.37	0.01
Mexico	mex	3.86	4.07	0.03	4.08	0.02
Rest of the Caribbean	xcb	3.58	3.92	0.07	2.23	0.04
Chile	chl	3.38	3.36	0.03	3.55	0.01
Argentina	arg	3.21	3.23	0.03	3.69	0.07
Latvia	lva	3.19	5.00	0.31	2.78	0.03
Lithuania	ltu	3.09	4.12	0.24	3.62	0.17
Turkey	tur	2.96	2.86	0.07	2.72	0.01
Thailand	tha	2.85	2.22	0.02	2.68	0.01
China	chn	2.59	2.12	0.01	2.29	0.00
Tunisia	tun	2.53	2.46	0.04	2.26	0.04
Rest of North Africa	xnf	2.46	2.43	0.02	2.82	0.05

continued

Table A.7 Cont'd

Region	Code	p.c. E_k^L	p.c. E_k^U	ϵ	p.c. E_k^D	ϵ
Brazil	bra	1.86	1.87	0.02	1.80	0.01
Botswana	bwa	1.62	3.11	0.16	2.01	0.05
Rest of Andean Pact	xap	1.58	1.73	0.01	1.58	0.01
Uruguay	ury	1.53	2.37	0.04	1.85	0.04
Colombia	col	1.44	1.44	0.01	1.69	0.01
Indonesia	idn	1.43	1.13	0.00	1.49	0.01
Morocco	mar	1.31	1.47	0.02	1.15	0.01
Central America	xca	1.09	1.57	0.03	1.02	0.01
Zimbabwe	zwe	1.05	1.15	0.01	0.95	0.00
Albania	alb	1.05	1.62	0.05	0.84	0.00
Peru	per	1.04	1.23	0.02	1.12	0.00
India	ind	0.99	0.94	0.00	0.87	0.00
Philippines	phl	0.95	1.07	0.03	0.75	0.00
Rest of Oceania	xoc	0.77	1.10	0.02	0.89	0.01
Vietnam	vnm	0.75	0.89	0.04	0.46	0.02
Rest of South America	xsm	0.74	1.03	0.01	1.18	0.03
Rest of South African CU	xsc	0.70	1.44	0.08	0.81	0.01
Sri Lanka	lka	0.60	0.81	0.02	0.58	0.01
Rest of South Asia	xsa	0.56	0.61	0.00	0.46	0.00
Rest of Sub-Saharan Africa	xss	0.39	0.44	0.01	0.43	0.01
Rest of SADC	xsd	0.29	0.32	0.01	0.42	0.02
Bangladesh	bgd	0.28	0.37	0.01	0.24	0.00
Rest of South east Asia	xse	0.24	0.35	0.01	0.39	0.00
Zambia	zmb	0.20	0.34	0.01	0.30	0.03
Madagascar	mdg	0.15	0.22	0.01	0.13	0.00
Uganda	uga	0.14	0.21	0.00	0.11	0.00
Tanzania	tza	0.10	0.18	0.01	0.11	0.00
Mozambique	moz	0.09	0.20	0.01	0.17	0.04
Malawi	mwi	0.09	0.17	0.00	0.09	0.00

Table A.8 reports per capita upstream and downstream carbon trade balances (p.c. b_k^U and p.c. b_k^D) for every GTAP region k. (tonCO$_2$ per capita.) Regions are ordered by decreasing distance to the origin.

Table A.9 reports absolute and per capita total carbon responsibility (E_k and p.c. E_k), and the difference between total carbon responsibility and direct emissions on a per capita basis ($E_k - E_k^L$ p.c.) for every GTAP region k. (tonCO$_2$ and tonCO$_2$ per capita.)

Table A.8 Upstream and downstream carbon trade balances for 87 GTAP regions (tonCO$_2$ per capita)

Region	p.c. b_k^U	p.c. b_k^D	Region	p.c. b_k^U	p.c. b_k^D
Rest of North America	13.409	−5.101	Greece	0.582	−0.686
Hong Kong	12.341	7.218	Uruguay	0.841	0.320
Rest of EFTA	0.405	12.537	Croatia	0.729	−0.380
Switzerland	7.156	6.271	Spain	0.804	−0.164
Luxembourg	8.819	−2.043	Rest of South African CU	0.737	0.106
Singapore	7.535	1.318	Thailand	−0.627	−0.171
Malta	3.842	−1.731	Rest of East Asia	−0.092	−0.619
Estonia	−0.793	−4.016	Albania	0.573	−0.203
Ireland	−0.216	3.882	Korea	0.183	−0.552
Cyprus	2.696	−2.507	Rest of FTAA	0.018	−0.576
Australia	−2.546	2.407	China	−0.471	−0.302
Czech Republic	−1.999	−2.876	Rest of South America	0.288	0.441
Finland	−3.390	0.817	Slovenia	0.471	−0.118
Austria	2.846	1.898	Central America	0.480	−0.072
Sweden	1.988	2.672	Argentina	0.016	0.477
South Africa	−2.831	−0.261	Rest of North Africa	−0.028	0.359
United Kingdom	2.152	0.961	Rest of Oceania	0.327	0.124
United States	2.334	−0.320	Vietnam	0.136	−0.292
Russian Federation	−2.144	0.535	Indonesia	−0.303	0.054
Canada	−1.596	1.524	Mexico	0.201	0.212
Malaysia	−1.981	0.943	Tunisia	−0.072	−0.269
Denmark	0.955	1.872	Turkey	−0.109	−0.247
Venezuela	−1.143	1.712	Colombia	−0.008	0.246
France	1.644	1.222	Philippines	0.121	−0.203
Japan	1.703	1.065	Morocco	0.162	−0.157
Germany	1.451	1.297	Sri Lanka	0.208	−0.019
Latvia	1.808	−0.419	Peru	0.191	0.077
Taiwan	−1.539	−0.973	Rest of South-east Asia	0.113	0.148
Hungary	−0.443	−1.598	Chile	−0.019	0.174
Belgium	1.567	0.267	Zambia	0.138	0.100
Slovakia	−0.229	−1.539	Rest of Andean Pact	0.145	0.000
Botswana	1.490	0.386	Zimbabwe	0.097	−0.101
Portugal	1.375	−0.573	Mozambique	0.108	0.085
R. of F. Soviet Union	−0.858	−1.187	Rest of SADC	0.032	0.133
Bulgaria	−0.662	−1.263	India	−0.052	−0.120
Rest of the Caribbean	0.340	−1.351	Rest of South Asia	0.050	−0.099
The Netherlands	1.343	−0.160	Bangladesh	0.088	−0.039
New Zealand	0.603	1.123	Malawi	0.082	0.004
Rest of Europe	0.112	−1.259	Tanzania	0.079	0.001
Rest of Middle East	−0.902	0.881	Uganda	0.073	−0.024
Lithuania	1.035	0.530	Madagascar	0.072	−0.016
Italy	1.115	0.091	Brazil	0.006	−0.066
Poland	−0.555	−0.837	R. of Sub-Saharan Africa	0.048	0.043
Romania	−0.322	−0.871			

Table A.9 Absolute and per capita total carbon responsibility and the difference
between total carbon responsibility and direct emissions on a per capita
basis for 87 GTAP regions ($MtonCO_2$ and $tonCO_2$ per capita)

Region	Code	E_k	p.c. E_k	p.c. $(E_k - E_k^L)$
United States	usa	6286.36	22.65	1.01
China	chn	2798.08	2.20	−0.39
Japan	jpn	1466.43	11.56	1.38
Russian Federation	rus	1385.65	9.51	−0.80
Rest of Middle East	xme	1198.06	6.90	−0.01
Germany	deu	1004.87	12.26	1.37
India	ind	936.18	0.91	−0.09
United Kingdom	gbr	710.81	12.00	1.56
France	fra	595.22	10.00	1.43
Rest of Former Soviet Union	xsu	573.07	4.16	−1.02
Canada	can	546.54	17.50	−0.04
Italy	ita	509.77	8.86	0.60
Mexico	mex	410.78	4.07	0.21
Korea	kor	388.93	8.17	−0.18
Australia	aus	350.21	18.03	−0.07
Spain	esp	318.33	8.07	0.32
Brazil	bra	315.80	1.83	−0.03
Poland	pol	282.85	7.30	−0.70
Indonesia	idn	278.76	1.31	−0.12
Rest of North Africa	xnf	265.27	2.63	0.17
South Africa	zaf	256.60	5.92	−1.55
Taiwan	twn	219.90	9.86	−1.26
The Netherlands	nld	213.97	13.40	0.59
Rest of Sub-Saharan Africa	xss	189.15	0.44	0.05
Turkey	tur	184.50	2.79	−0.18
Venezuela	ven	162.82	6.60	0.28
Thailand	tha	153.63	2.45	−0.40
Belgium	bel	135.87	13.21	0.92
Argentina	arg	129.66	3.46	0.25
Hong Kong	hkg	124.37	17.36	9.78
Malaysia	mys	104.49	4.41	−0.52
Rest of South Asia	xsa	104.12	0.54	−0.02
Greece	grc	100.81	9.54	−0.05
Switzerland	che	99.42	13.88	6.71
Rest of FTAA	xfa	99.35	3.83	−0.28
Rest of East Asia	xea	92.94	3.64	−0.36
Czech Republic	cze	92.44	9.04	−2.44
Rest of EFTA	xef	88.59	18.35	6.47
Finland	fin	88.18	16.99	−1.29
Austria	aut	86.43	10.67	2.37
Denmark	dnk	82.32	15.42	1.41
Sweden	swe	80.38	9.06	2.33
Romania	rom	80.31	3.59	−0.60

Table A.9 Cont'd

Region	Code	E_k	p.c. E_k	p.c. $(E_k - E_k^L)$
Portugal	prt	74.02	7.40	0.40
Philippines	phl	72.72	0.91	−0.04
Hungary	hun	71.97	7.20	−1.02
Colombia	col	67.25	1.56	0.12
Rest of Europe	xer	64.34	3.79	−0.57
Singapore	sgp	63.67	19.12	4.43
Vietnam	vnm	53.54	0.67	−0.08
Chile	chl	53.14	3.46	0.08
Ireland	irl	51.26	13.58	1.83
Central America	xca	47.75	1.30	0.20
Bangladesh	bgd	39.96	0.30	0.02
Rest of the Caribbean	xcb	38.47	3.08	−0.51
Morocco	mar	38.31	1.31	0.00
Rest of Andean Pact	xap	35.34	1.65	0.07
Bulgaria	bgr	33.20	4.09	−0.96
Slovakia	svk	32.27	5.96	−0.88
New Zealand	nzl	30.67	7.97	0.86
Peru	per	30.66	1.17	0.13
Rest of SADC	xsd	25.30	0.37	0.08
Rest of South-east Asia	xse	24.80	0.37	0.13
Tunisia	tun	22.85	2.36	−0.17
Croatia	hrv	20.85	4.71	0.17
Slovenia	svn	18.79	9.43	0.18
Estonia	est	14.27	9.99	−2.40
Lithuania	ltu	14.25	3.87	0.78
Sri Lanka	lka	13.41	0.69	0.09
Zimbabwe	zwe	12.91	1.05	0.00
Luxembourg	lux	12.56	28.49	3.39
Latvia	lva	9.27	3.89	0.69
Cyprus	cyp	8.28	10.88	0.09
Rest of Oceania	xoc	8.05	1.00	0.23
Rest of South America	xsm	7.86	1.11	0.36
Uruguay	ury	7.09	2.11	0.58
Rest of South African CU	xsc	5.69	1.13	0.42
Tanzania	tza	4.97	0.14	0.04
Albania	alb	4.24	1.23	0.19
Botswana	bwa	4.17	2.56	0.94
Uganda	uga	3.64	0.16	0.02
Mozambique	moz	3.34	0.19	0.10
Zambia	zmb	3.27	0.32	0.12
Malta	mlt	2.87	7.27	1.06
Madagascar	mdg	2.79	0.17	0.03
Rest of North America	xna	2.42	19.22	4.15
Malawi	mwi	1.47	0.13	0.04

Bibliography

F. Ackerman, M. Ishikawa and M. Suga. The carbon content of Japan-US trade. *Energy Policy*, 35: 4455–62, 2007.

N. Ahmad and A. W. Wyckoff. Carbon dioxide emissions embodied in international trade of goods. STI Working Paper DSTI/DOC 15, Organisation for Economic Co-operation and Development (OECD), Paris, 2003.

C. Albin. Negotiating international cooperation: Global public goods and fairness. *Review of International Studies*, 29(3): 365–85, 2003.

R. Andrew, G. P. Peters and J. Lennox. Approximation and regional aggregation in multi-regional input–output analysis. *Economic Systems Research*, 21: 311–335, 2009.

Association of British Insurers (ABI). *Risk Returns and Responsibility*. Association of British Insurers, London, 2004. (http://www.abi.org.uk).

T. Azomahou, F. Laisney and P. Nguyen Van Economic development and CO_2 emissions: A nonparametric panel approach. *Journal of Public Economics*, 90(6-7): 1347–63, 2006.

S. Barrett. Self enforcing international environmental agreements. *Oxford Economic Papers*, 46: 878–94, 1994.

S. Bastianoni, F. M. Pulselli and E. Tiezzi. The problem of assigning responsibility for greenhouse gas emissions. *Ecological Economics*, 49: 253–7, 2004.

D. Bazin, J. Ballet and D. Touahri. Environmental responsibility versus taxation. *Ecological Economics*, 49: 129–34, 2004.

British Standards Institution (BSI). *PAS 2050 – Specification for the Assessment of the Life Cycle Greenhouse Gas Emissions of Goods and Services*. BSI British Standards, London, 2008. (http://www.bsi-global.com/en/Standards-and-Publications/How-we-can-help-you/Professional-Standards-Service/PAS-2050/).

M. Brockmeier. A graphical exposition of the GTAP model. GTAP Technical Paper 8. Center for Global Trade Analysis, Purdue University, West Lafayette, IN, 2001.

J. G. Canadell, C. Le Quere, M. R. Raupach, C. B. Field, E. T. Buitenhuis, P. Ciais, T. J. Conway, N. P. Gillett, R. A. Houghton and G. Marland. Contributions to accelerating atmospheric CO_2 growth from economic activity. *Proceedings of the National Academy of Sciences*, 104(47): 18866–70, 2007.

C. Carraro and D. Siniscalco. Strategies for the international protection of the environment. *Journal of Public Economics*, 52: 309–28, 2003.

P. Cerin. Bringing economic opportunity into line with environmental influence: A discussion on the Coase theorem and the Porter and van der Linde hypothesis. *Ecological Economics*, 56: 209–25, 2006.

G. Chichilnisky and G. Heal. Who should abate carbon emissions? An international perspective. *Economic Letters*, 44: 443–49, 1994.

B. R. Copeland and M. S. Taylor. Free trade and global warming: A trade theory view of the Kyoto protocol. *Journal of Environmental Economics and Management*, 49: 205–34, 2005.

E. Davar. Input–output and general equilibrium. *Economic Systems Research*, 1(3): 331–43, 1989.

E. Dietzenbacher. In vindication of the Ghosh model: A reinterpretation as a price model. *Journal of Regional Science*, 37(4): 629–51, 1997.

B. V. Dimaranan (ed.) *Global Trade, Assistance and Production: The GTAP 6 Data Base*. Center for Global Trade Analysis, Purdue University, West Lafayette, IN, 2006.

F. Duchin. A world trade model based on comparative advantage with m regions, n goods and k factors. *Economic Systems Research*, 17: 141–62, 2005.

U. Ebert and A. Welsch. Meaningful environmental indices: A social choice approach. *Journal of Environmental Economics and Management*, 47: 270–83, 2004.

The Economist. Green.view: On the mark. *Economist*, (28 January), 2008. (http://www.economist.com/world/international/displaystory.cfm? story _id=10563590).

P. Eder and M. Nadoroslawsky. What environmental pressures are a regions' industries responsible for? A method of analysis with descriptive indices and input–output models. *Ecological Economics*, 29: 359–74, 1999.

P. Ekins and T. Barker. Carbon taxes and carbon emissions trading. *Journal of Economic Surveys*, 15(3): 325–76, 2001.

A. D. Ellerman. A note on tradeable permits. *Environmental and Resource Economics*, 31: 123–31, 2005.

European Platform on Life Cycle Assessment (EPLCA). *Carbon Footprint – What it is and How to Measure it*. EPLCA, Ispra, Italy, 2008. (http://lca.jrc.ec.europa.eu/).

EUROSTAT. *ESA 95 Supply, Use and Input–Output Tables*. Statistical Office of the European Communities, Brussels, 2009. (http://epp.eurostat.ec.europa.eu/portal/page?_pageid=2474,54156821,2474_54764840&_dad=portal&_schema=PORTAL).

J-J. Ferng. Allocating the responsibility of CO_2 over-emissions from the perspectives of benefit principle and ecological deficit. *Ecological Economics*, 46: 121–41, 2003.

B. Gallego and M. Lenzen. A consistent input–output formulation of shared producer and consumer responsibility. *Economic Systems Research*, 17(4): 365–91, 2005.

A. Ghosh. Input–output approach in an allocation system. *Economica*, 25: 58–64, 1958.

S. Giljum and K. Hubacek. Alternative approaches of physical input–output analysis to estimate primary material inputs of production and consumption activities. *Economic Systems Research*, 16(3): 301–10, 2004.

S. Giljum, C. Lutz and A. Jungnitz. The global resource accounting model (GRAM): A methodological concept paper. SERI Working Paper 8, Sustainable Europe Research Institute, Vienna, 2008.

M. Grubb and K. Neuhoff. Allocation and competitiveness in the EU emissions trading scheme: Policy overview. *Climate Policy*, 6(1): 7–30, 2006.

N. Hanley, J. Shogren and B. White. *Environmental Economics in Theory and Practice*. Palgrave, London, 2007.

J. Hansen and M. Sato. Greenhouse gas growth rates. *Proceedings of the National Academy of Sciences*, 101: 16109–14, 2004.

G. Heal. Corporate social responsibility: An economic and financial framework. *The Geneva Papers*, 30(3): 387–409, 2005.

E. G. Hertwich. Life cycle approaches to sustainable consumption: A critical review. *Environmental Science and Technology*, 39(13): 4673–84, 2005.

E. G. Hertwich and G. P. Peters. Carbon footprint of nations: A global, trade-linked analysis. *Environmental Science and Technology*, 43: 6414–6420, 2009.

R. Hoekstra and M. A. Janssen. Environmental responsibility and policy in a two-country dynamic input–output model. *Economic Systems Research*, 18(1): 61–84, 2006.

R. Hoekstra and J. C. J. M. van den Bergh. Constructing physical input–output tables for environmental modeling and accounting: Framework and illustrations. *Ecological Economics*, 59(3): 375–93, 2006.

M. Hoel and K. Schneider. Incentives to participate in an international environmental agreement. *Environmental and Resource Economics*, 9: 153–70, 1997.

Intergovernmental Panel on Climate Change (IPCC). *Climate Change 2007: The Physical Science Basis. Contribution of Working Group I to the Fourth Assessment Report of the Intergovernmental Panel on Climate Change.* Cambridge University Press, Cambridge, 2007.

International Energy Agency (IEA). *Key World Energy Statistics 2003.* IEA, Paris, 2003.

International Energy Agency (IEA). *Energy Balances of Non-OECD Countries.* IEA/OECD, Paris, 2004.

International Organisation for Standardisation (ISO). *ISO 14040: Environmental Management – Life Cycle Assessment – Principles and Framework.* ISO, Geneva, 2006.

J. Jesper and T. N. Rasmussen. Allocation of CO_2 emissions permits: A general equilibrium analysis of policy instruments. *Journal of Environmental Economics and Management,* 40: 111–36, 2000.

M. Kainuma, Y. Matsuoka and T. Morita. Estimation of embodied CO_2 emissions by general equilibrium model. *European Journal of Operational Research,* 122(2): 392–404, 2000.

Y. Kondo, Y. Moriguchi and H. Shimizu. CO_2 emissions in Japan: Influences of imports and exports. *Applied Energy,* 59(2–3): 163–74, 1998.

J. Kornai. Resource-constrained versus demand-constrained systems. *Econometrica,* 47(4): 801–20, 1979.

A. Lange. The impact of equity-preferences on the stability of international environmental agreements. *Journal of Environmental Economics and Management,* 34: 247–67, 2006.

H-L. Lee. An emissions data base for integrated assessment of climate change policy using GTAP. GTAP Resource 1143, Global Trade Analysis Project, Purdue University, West Lafayette, IN, 2007. (https://www.gtap. agecon.purdue.edu/resources/res_display.asp?RecordID=1143).

M. Lenzen. Errors in conventional and input–output-based life-cycle inventories. *Journal of Industrial Ecology,* 4(4): 127–48, 2001.

M. Lenzen, L-L. Pade and J. Munksgaard. CO_2 multipliers in multi-region input–output models. *Economic Systems Research,* 16(4): 391–412, 2004.

M. Lenzen, J. Murray, F. Sack and T. Wiedmann. Shared producer and consumer responsibility – theory and practice. *Ecological Economics,* 61(1): 27–42, 2007.

W. Leontief and D. Ford. Environmental repercussions and the economic structure: An input–output approach. *Review of Economics and Statistics,* 52: 262–71, 1970.

C. Lutz, B. Meyer and M. I. Wolter. GINFORS-model. MOSUS Workshop, IIASA Laxenburg; 14–15, April, Gesellschaft fr Wirtschaftliche Strukturforschung mbH (GWS), Osnabrück, 2005.

S. McDonald and K. Thierfelder. Deriving a global social accounting matrix from GTAP versions 5 and 6 data. GTAP Technical Paper 22, Center for Global Trade Analysis, Purdue University, West Lafayette, IN, 2004.

S. Mandell. Optimal mix of emissions taxes and cap-and-trade. *Journal of Environmental Economics and Management*, 56(2): 131–40, 2008.

M. Meharara. Energy consumption and economic growth: The case of oil exporting countries. *Energy Policy*, 35: 2939–45, 2007.

R. E. Miller and P. D. Blair. *Input–Output Analysis: Foundations and Extensions*. Prentice-Hall, Englewood Cliffs, NJ, 1985.

B. Muller. Food miles or poverty eradication? Oxford energy and environment comment, Oxford Institute for Energy Studies, University of Oxford, Oxford, 2007.

J. Munksgaard and K. A. Pedersen. CO_2 accounts for open economies: Producer or consumer responsibility? *Energy Policy*, 29: 327–34, 2001.

J. Murray and C. Dey. Carbon neutral – sense and sensibility. ISA Research Paper 07/02, Centre for Integrated Sustainability Analysis, University of Sydney, Sydney, 2007. (http://www.isa.org.usyd.edu.au/publications/CarbonNeutral.pdf).

O. K. Olafsson. Iceland and Norway join race for carbon neutrality. *IceNews*, (8 April), 2008. (http://www.icenews.is/index.php/tag/carbon-neutral/).

J. Oosterhaven. Leontief versus Ghoshian price and quantity models. *Southern Economic Journal*, 62: 750–9, 1996.

J. Oosterhaven, D. Stelder and S. Inomata. Evaluation of non-survey international IO construction methods with the Asian-Pacific input–output table. IDE Discussion Papers 114, Institute of Developing Economies, Japan External Trade Organization (JETRO), Tokyo, Japan, 2007.

T. Panayotou, J. D. Sachs and A. P. Zwane. Compensation for meaningful participation in climate change control: A modest proposal and empirical analysis. *Journal of Environmental Economics and Management*, 43: 437–54, 2002.

G. P. Peters and E. G. Hertwich. Structural analysis of international trade: environmental impacts of Norway. *Economic Systems Research*, 18: 155–81, 2006.

G. P. Peters and E. G. Hertwich. CO_2 embodied in international trade with implications for climate change policy. *Environmental Science and Technology*, 42(5): 1401–07, 2008a.

G. P. Peters and E. G. Hertwich. Post-Kyoto greenhouse gas inventories: Production versus consumption. *Climatic Change*, 86: 51–66, 2008b.

J. C. V. Pezzey. Emission taxes and tradeable permits: A comparison of views on long-term efficiency. *Environmental and Resource Economics*, 26: 329–42, 2003.

W. A. Pizer. Combining price and quantity controls to mitigate global climate change. *Journal of Public Economics*, 85(3): 409–34, 2002.

J. Proops, M. Faber and G. Wagenhals. *Reducing CO_2 Emissions: A Comparative Input–Output Study for Germany and the UK*. Springer, Berlin, 1993.

J. Proops, G. Atkinson, B. F. v. Schlotheim and S. Simon. International trade and the sustainability footprint: A practical criterion for its assessment. *Ecological Economics*, 28: 75–97, 1999.

T. ten Raa. *The Economics of Input–Output Analysis*. Cambridge University Press, Cambridge, 2006.

J. Rawls. *A Theory of Justice*. Belknap Press, Cambridge, MA, 1971.

J. Rodrigues and T. Domingos. Environmental responsibility: Comparing two approaches. *Ecological Economics*, 66: 533–46, 2008.

J. Rodrigues and S. Giljum. The accounting of indirect material requirements in material flow-based indicators. *The ICFAI Journal of Environmental Economics*, III(2): 51–69, 2005.

J. Rodrigues, T. Domingos, S. Giljum and F. Schneider. Designing an indicator of environmental responsibility. *Ecological Economics*, 59(3): 256–66, 2006.

K-M. Schulte. Scientific consensus on climate change? *Energy and Environment*, 19(2): 281–6, 2008.

U. Srinivasan, S. Carey, E. Hallstein, P. Higgins, A. Kerr, L. Koteen, A. Smith, R. Watson, J. Harte and R. Norgaard. The debt of nations and the distribution of ecological impacts from human activities. *Proceedings of the National Academy of Sciences*, 105(5): 1768–73, 2008.

A. Tukker and B. Jansen. Environmental impacts of products: a detailed review of studies. *Journal of Industrial Ecology*, 10(3): 159–182, 2006.

A. Tukker, E. Poliakov, R. Heijungs, T. Hawkins, F. Neuwahl, J. Rueda-Cantuche, S. Giljum, S. Moll, J. Oosterhaven and M. Bouwmeester. Towards a global multi-regional environmentally extended input–output database. *Ecological Economics*, 68(7): 1928–37, 2009.

United Nations (UN). Handbook of input–output compilation and analysis. Studies in Methods ST/ESA/STAT/SER.F/74, United Nations, New York, NY, 1999.

United Nations (UN). System of National Accounts (SNA) 1993. Studies in Methods Series F., No. 2, Rev. 4, Sales No. 94.XVII.4, United Nations, New York, 1994.

United Nations Framework Convention on Climate Change (UNFCCC). *Kyoto Protocol to the United Nations Framework Convention on Climate Change*. Bonn, 1998. (http://unfccc.int/resource/docs/convkp/kpeng.pdf).

United Nations Framework Convention on Climate Change (UNFCCC). *Key GHG Data: Greenhouse Gas (GHG) Emissions Data for 1990–2003 submitted to the UNFCCC.* Bonn, 2005. (http://unfccc.int/resource/docs/publications/key_ghg.pdf).

C. L. Weber and H. S. Matthews. Embodied environmental emissions in US international trade. *Environmental Science and Technology*, 41: 4875–81, 2007.

H. Weisz and F. Duchin. Physical and monetary input–output analysis: What makes a difference? *Ecological Economics*, 57(3): 534–41, 2006.

T. Wiedmann, M. Lenzen and R. Wood. *Uncertainty Analysis of the UK-MRIO Model – Results from a Monte-Carlo Analysis of the UK Multi- Region Input–Output Model (Embedded Emissions Indicator).* Department for Environment, Food and Rural Affairs (DEFRA), London, 2008a.

T. Wiedmann, M. Lenzen, K. Turner and J. Barrett. Examining the global environmental impact of regional consumption activities, Part 2: Review of input–output models for the assessment of environmental impacts embodied in trade. *Ecological Economics*, 61: 15–26, 2007.

T. Wiedmann, R. Wood, M. Lenzen, J. Minx, D. Guan and J. Barrett. *Development of an Embedded Carbon Emissions Indicator – Producing a Time Series of Input–Output Tables and Embedded Carbon Dioxide Emissions for the UK by Using a MRIO Data Optimisation System.* Department for Environment, Food and Rural Affairs (DEFRA), London, 2008b.

World Business Council for Sustainable Development (WBCSB). *The Greenhouse Gas Protocol: A Corporate Accounting and Reporting Standard.* Geneva, 2004. Rev. ed.

Worldwatch Institute. *Biofuels for Transport: Global Potential and Implications for Agriculture.* Earthscan, London, 2007.

N. Yamano and N. Ahmad. The OECD input–output database: 2006 edition. STI Working Paper DSTI/DOC(2006)8, OECD, Paris, 2006.

Index

Agent (economic) 10
Attributed CO_2 emissions 23

Benefit principle and ecological deficit 25

Carbon emissions added (CEA) 26
Carbon emissions direct 7; indirect 9
Carbon intensity 8
Carbon responsibility (total) 33, 46; of consumption 34, 44; of production 34, 44
Carbon trade balance (downstream DCTB, and upstream, UCTB) 70
CO_2 trade balance 24
Consumption 11

Direct (or local) carbon emissions of a sector 9
Domestic indirect emissions 53

Economic causality (property) 38
Economic flow 9, 11
Elasticity of emissions 61
Emissions embodied in trade 28
Environmental indicator 9
Error estimation 53; margin 64

Ghosh inverse 18
Global Trade, Assistance and Production (GTAP) Database 48
Greenhouse gas (GHG) 2
Greenhouse warming potential (GWP) 9
Gross domestic product (GDP) 60

Input–output (IO) analysis 6; environmental 7; multi-regional (MRIO) 47
International indirect emissions 53
International Panel on Climate Change (IPCC) 1

Kyoto Protocol 1

Leontief, W. 6
Leontief inverse 18
Life cycle assessment (LCA) 20

Monotonicity (property) 36

Normalization (property) 35

Price (market, agent or world) 49
Probability distribution 57
Production 11

RAS method 53
Rawls, J. 33
Region 10; GTAP 48; aggregated world 71

Scale invariance (property) 35
Sector 10; internal and external 9
Shared responsibility 27
Social accounting matrix (SAM) 8
Symmetry (property) 39
System of National Accounts (SNA) 8

Total downstream carbon intensity of a flow (TDCIF) 16; of a sector (TDCIS) 16

Total downstream embodied carbon
emissions: of a flow (TDECEF) 16;
of a sector (TDECES) 16
Total indirect effects (property) 38
Total upstream carbon intensity of a
flow (TUCIF) 13; of a sector
(TUCIS) 14

Total upstream embodied carbon
emissions of a flow (TUECEF) 12; of
a sector (TUECES) 13
Trade share method 52

United Nations Framework Convention
on Climate Change (UNFCCC) 1